Organic Winegrowing Manual

Technical Editor
Glenn T. McGourty

Publication Coordinators
Jeri Ohmart and David Chaney

2011

University *of* **California** Agriculture and Natural Resources

Funding for writing this publication provided in part by the California
Department of Food and Agriculture *Buy California Initiative.*

CALIFORNIA DEPARTMENT OF
FOOD AND AGRICULTURE

To order or obtain ANR publications and other products, visit the ANR Communication Services online catalog at http://anrcatalog.ucdavis.edu or phone 1-800-994-8849. You can also place orders by mail or FAX, or request a printed catalog of our products from

University of California
Agriculture and Natural Resources
Communication Services
1301 S. 46th Street
Building 478 - MC 3580
Richmond CA 94804-4600

Telephone 1-800-994-8849 or 510-665-2195
Fax: 510-665-3427
E-mail: anrcatalog@ucdavis.edu

Publication 3511
ISBN-13: 978-1-60107-563-5
Library of Congress Control Number: 2011935274

Cover photograph credits: Tom Liden (front and back covers).

To simplify information, trade names of products have been used. No endorsement of named or illustrated products is intended, nor is criticism implied of similar products that are not mentioned or illustrated.

UC PEER REVIEWED This publication has been anonymously peer reviewed for technical accuracy by University of California scientists and other qualified professionals. This review process was managed by ANR Associate Editors for Pomology, Viticulture, and Subtropical Horticulture James Stapleton and Ben Faber.

Printed in the United States of America on recycled paper.

5m-pr-10/11-WJC/RW

Contents

■ Preface and Acknowledgments

The *Organic Winegrowing Manual* provides detailed information for growers on how to grow winegrapes using organic methods and materials. The book addresses production issues, economics, weed and disease management, the process of conversion from conventional to organic culture, and organic certification and registration, each one an essential element in planning for success in a highly competitive marketplace.

This manual has been developed in part as an adaptation and expansion of publications on organic crop production published earlier by the University of California's Division of Agriculture and Natural Resources (UC ANR). This earlier series was an important resource for organic and conventional vegetable growers going through the early stages of implementing compliance with then-newly defined national organic standards. Information from those publications has been incorporated into this new manual, along with a considerable amount of new information targeted specifically to winegrape growers.

Other information resources related to organic winegrape production are also available from UC ANR, including *Grape Pest Management, Wine Grape Varieties in California,* and a variety of free, downloadable publications on individual grape pests and diseases. In addition, the UC Integrated Pest Management program offers *UC IPM Grape Pest Management Guidelines* on its website (http://www. ipm.ucdavis.edu/PMG). *Pest Notes,* produced by UC IPM mainly for landscapers and home gardeners, may also be of interest (http://www.ipm.ucdavis. edu/PDF/PESTNOTES). You will find additional information on winegrapes at UC Integrated Viticulture Online (http://ucanr.org/sites/invit).

The authors are grateful to the True North Foundation, which supported work on this project, and Fetzer Vineyards and Bonterra Vineyards for their generous support of numerous collaborative trials with UC Cooperative Extension and for their openness in discussing their farming experiences with the authors. In particular we thank Tom Piper, Dave Koball, and co-author and Fetzer employee Ann Thrupp.

Funding for writing this publication was provided by the Buy California Initiative of the California Department of Food and Agriculture and by the United States Department of Agriculture. The content of the publication does not necessarily reflect the views or policies of CDFA or USDA, nor does any mention of trade names, commercial products, or organizations imply any endorsement by CDFA or USDA. The authors would like to thank Jeri Ohmart of the University of California Sustainable Agriculture Research and Education Program and David Chaney, formerly of that same program for their assistance in coordinating the manuscripts and materials for this publication.

■ Authors

TECHNICAL EDITOR

Glenn T. McGourty, University of California Cooperative Extension, Mendocino and Lake Counties

CONTRIBUTING AUTHORS

Kendra Baumgartner, USDA–ARS, UC Davis, Department of Plant Pathology

Janet C. "Jenny" Broome, UC Cooperative Extension, Sacramento County

Robert L. Bugg, formerly with UC Sustainable Agriculture Research and Education Program, Davis

David Chaney, DEC Education Services, Corvallis, Oregon (formerly with UC Sustainable Agriculture Research and Education Program, Davis)

Pete Christensen, Viticulture Specialist Emeritus, UC Kearney Agricultural Center

Michael J. Costello, California State University, San Luis Obispo

Gregory A. Giusti, UC Cooperative Extension, Mendocino County

Ray Green, California Department of Food and Agriculture Organic Program

W. Douglas Gubler, UC Davis, Department of Plant Pathology

Susan P. Harrison, UC Davis, Department of Environmental Science and Policy

W. Thomas Lanini, UC Davis, Weed Science Program

Adina Merenlender, UC Berkeley, Department of Environmental Science, Policy, and Management

Tom Piper, Mendocino County Department of Agriculture

John Reganold, Washington State University

L. Ann Thrupp, Fetzer Vineyards, Hopland

PUBLICATION COORDINATORS

Jeri Ohmart, UC Sustainable Agriculture Research and Education Program, Davis

David Chaney, DEC Education Services, Corvallis, Oregon

PART 1

Introduction and Overview

PHOTO TOM LIDEN

PHOTO: TOM LIDEN

Introduction: Organic Winegrowing Farming Systems

GLENN T. McGOURTY, L. ANN THRUPP, AND JANET C. "JENNY" BROOME

Organic winegrowing has become an important sector in California's wine industry. In 2008, there were almost 10,000 acres of organically certified vineyards (CDFA 2008). Total global acreage of organic grapes has been estimated at just over 228,000 acres for 31 countries (including California acreage), with Italy alone accounting for 77,000 acres (Willer and Yussefi 2006). Organic winegrowing in California is mostly practiced in the coastal growing areas, where wine quality and pricing allow the use of more "hand-crafted" practices in the vineyard and in the winery. Pest and disease pressure are also lower there than in interior winegrape growing regions, where price constraints require growers and vintners to minimize labor inputs and maximize yields. At the time of this writing, organic winegrowing is not widely practiced in the interior valleys of California.

Winegrowers switch to organic practices for a variety of reasons. Some are growing for specific market niches. Others farm organically to address their own concerns about the long-term health of their property, workers, and families, since they believe that the elimination of conventional agricultural chemicals in their farming systems is environmentally more desirable. Others find that they can minimize their neighbors' health concerns by adopting organic farming practices, a factor that can be very important when a vineyard is in an urban/agricultural interface area.

The switch to organic winegrowing requires some changes in one's philosophy about farming.

Organic growing is not just about changing the inputs from conventional agrichemicals to other materials that are acceptable to organic certifiers. Organic winegrowers are encouraged to think of the farm as a mini-ecosystem that can self-regulate for pests, diseases, and fertility, rather than as a conventional agriculture system that requires the input of numerous off-farm materials (e.g., fertilizer, water, crop protectants) in order to produce marketable outputs (crops and animal products). In practice, organic winegrowers use no synthetically manufactured pesticides or fertilizers to produce winegrapes. Instead, they improve soil fertility and quality with composts, cover crops, and mined minerals, seeking to build soil organic matter as a source of biological resilience, stored energy, chemical buffering, and plant nutrients. You will frequently hear organic winegrowers say that they want to bring "life" into their farming system. Most organic growers view the farm with a "systems approach" in which the interconnected web of life in the vineyard is noted, appreciated, and counted on. They recognize that any one practice may greatly influence other elements in the vineyard, such as pest pressures or fruit quality. This ecological orientation differs from that of many conventional growers who do not rely on nature in the same way to maintain soil fertility and pest control, and choose instead to apply synthetic crop protectants and chemical fertilizers.

Like many conventional growers, organic winegrowers use integrated pest management (IPM) techniques to control diseases and pests.

IPM emphasizes field monitoring to anticipate problems and react quickly if intervention is needed. A primary IPM strategy is to monitor the crop area frequently for both insect pest and predator numbers. The grower's knowledge of pest and disease life cycles is critical, and action threshold levels are established to guide decisions about the timing and frequency of pest control measures. Many of the insecticides used in organic winegrowing have low toxicity and limited effectiveness and will not control pests if used by themselves.

Cultural practices are also important to an IPM program. For example, pulling off leaves from around fruit clusters just after flowering creates good air movement and improves sunlight infiltration, thus minimizing Botrytis bunch rot. Also, the exposed clusters are easier to cover with organic fungicides, which reduce powdery mildew problems. Finally, leaf pulling reduces insect pest pressure, since the first generation of leafhoppers feeds on these leaves. Without a food source, the pests perish as the removed leaves dry on the vineyard floor. This helps to reduce leafhopper numbers and the potential for infestations later in the growing season.

Organic winegrowers consciously try to provide habitat in their vineyards to attract beneficial insects, mites, and spiders in an effort to make spraying unnecessary. Fortunately, in many organic vineyards insect and mite pests often are not a problem once the growers stop spraying powerful, conventional, broad-spectrum insecticides. Generalist predator and parasitoid insect populations build rapidly in many vineyards, especially in coastal regions, following the cessation of broad-spectrum insecticide applications. Habitat modification may be as simple as managing the naturalized weeds in the vineyard, or it may be more elaborate, involving the planting of selected insectary plants including cover crops, hedgerows, "island plantings," or other arrangements that encourage a resident population of beneficial insects, spiders, and mites.

Powdery mildew control is accomplished with sulfur sprays or dust and other effective, naturally occurring materials. Many organic winegrowers rely on electronic weather monitoring and the powdery mildew Risk Assessment Index or RAI (also known as the Gubler-Thomas model), developed by University of California scientists to optimize fungicide use and disease control (see also chapter 8, Organic Grapevine Disease Management in California). By using the RAI with field monitoring, growers can precisely time the application of fungicides to ensure that they are getting good crop protection without making excessive applications.

As organic matter is added to the soil, diverse organisms in the root zone change the dynamics of soilborne pathogen activity and the pathogens' ability to cause disease on the vine roots. Organically farmed vineyards infested with phylloxera have been shown to last many years longer than conventionally farmed vineyards attacked by phylloxera, although they do need to be replanted eventually (Lotter, Granett, and Omer 1999). This increased vine longevity can be attributed to a diverse microflora in the soil that suppresses the pathogenic fungi that attack grapevine roots damaged by phylloxera.

Weed control in an organic vineyard is done mostly by specialized equipment that tills beneath the vine rows. Vine middles are mowed, disked, or otherwise tilled. Selected cover crops can also be planted to compete with specific weeds under certain conditions. This is perhaps the greatest challenge that organic winegrowers face in the transition from conventional to organic farming: Mechanical tillage weed control equipment is expensive and requires patience and skill to maintain and operate. With high fuel costs, mechanical tillage and weed control is generally more expensive than chemical weed control, particularly if the vineyard requires additional hand hoeing. Some organic winegrowers also use a propane burner to flame-kill their weeds when they are very small seedlings, usually in the late fall and early winter. Organic winegrowers also rely on hand labor to either mow or hoe any weeds that escape tillage. While organically acceptable herbicides are available, most are very expensive and not highly effective.

Many organic winegrowers have a strong conservation ethic and go beyond just the minimum requirements for organic certification. For example, many invite nature and biodiversity into their vineyards by providing habitat fixtures such as perches or houses for raptors and songbirds and by protecting and enhancing riparian areas along creeks and rivers. Fencing is often designed to exclude deer from their vineyards while allowing wildlife access to these ecologically productive and important riparian areas.

Since many organic winegrowers live in their vineyards, they want a place that is safe for their family and workers, free of toxic chemicals that might cause unintended pesticide exposure or pose other health risks. In addition, many of their nonfarming neighbors appreciate their dedication to a farming system that values nature and works with ecological processes rather than synthetic pesticides to control pests. There are many wineries and vineyards in the urban/rural interface, and being perceived as a "good neighbor" can make farming easier in an environment where every practice is likely to be scrutinized by adjacent property owners.

Finally, wine made from organically grown grapes can be of the highest quality and show the best flavors that a vineyard can deliver for the region. The organic winegrowing movement in California was inspired by the high quality and flavors of fresh organic vegetables offered by local market gardeners and prepared by creative chefs in gourmet restaurants around California during the 1980s. The early organic winegrowers were similarly inspired to achieve more flavor intensity, high quality, and a distinctive product in their winegrapes and wine.

THREE STYLES OF ORGANIC WINEGROWING

In California, there are three different styles of organic winegrowing:

Heritage organic winegrowing. This method uses traditional practices that have been employed since winegrowing began in California over 150 years ago. The heritage approach is characterized by vineyards in which vines are head pruned. Most of these vineyards are not irrigated, so the vineyard floor is cultivated during the growing season to conserve moisture. Planted cover crops are mostly winter annuals such as mustards, radishes, and mixes of small grains and legumes. Yields are low to moderate (2 to 4 tons per acre). Sulfur dust is used to control powdery mildew. Under normal conditions, even organically approved insecticides and miticides are not used. Typical cultivars planted in this system include Zinfandel, Petite Sirah, Carignane, Mourvedre, and Alicante Bouschet. St. George rootstock is common, although some isolated vineyards are grown on their own roots. Compost, if needed, is applied annually at the rate of 1 to 2 tons per acre. Many of these vineyards are very old. Frequently, they are on upland sites on soils with moderate water-holding capacity and fertility. In the best vineyards, winemakers hotly compete for the opportunity to purchase fruit, as the concentration of color and flavors and the intensity of the resulting wine can be very impressive. It is not expensive to farm these vineyards, since pruning and canopy management operations are fairly simple. Because yields are low, though, the potential for a large income from the vineyard is limited. A small number of vineyards (around 500 acres) are farmed this way in California. Growers who use this approach tend to be conservative in their farming techniques, and this is not considered to be a management-intensive farming system. Many management decisions are made based on the calendar and soil moisture.

Modern organic winegrowing. A second approach, modern organic winegrowing follows more conventional practices including wire trellises, drip and sprinkler irrigation, international varieties (Cabernet Sauvignon, Chardonnay, Sauvignon Blanc, Pinot Noir, and Merlot), and more intensive canopy management. Many of these vineyards are planted on fertile soils, and yields may range from 4 to 8 tons per acre depending on the cultivar. Vineyard floors are managed in a wide array of styles, from nontillage planted to perennial grasses, to tilled and annually planted grass and legume mixes. Vineyards are closely monitored and organically approved materials are often sprayed as crop protectants to prevent disease and vine pests from causing economic losses. Canopy management has two roles in the vineyard: It is a critical part of an IPM program, and it helps to improve fruit quality. Compost and concentrated organic fertilizers are often used to keep vegetative vine growth balanced with the large crops that often set, mostly for white cultivars. Beneath-the-vine weed control is carried out mostly by machines, but some growers also use flaming. Management decisions require careful attention to growing conditions, and this style of farming is management intensive. Expenses are similar to those incurred by conventional growers. Yields and economic returns are also similar to those of conventional winegrowing systems (Smith et al. 2004; Weber, Klonsky, and DeMoura 2005). Most of California's 10,000 acres of organically certified vineyards are farmed this way.

Biodynamic winegrowing. Biodynamic winegrowing is similar to modern organic winegrowing but emphasizes even more intensely the value of managing the vineyard as a mini-ecosystem. There is a strong desire by many biodynamic winegrowers to look internally, within their vineyard, to solve pest management and soil fertility problems. Biodynamic winegrowers try to minimize off-farm inputs into their vineyards by creating as closed a farming system as possible. Composting is considered key to soil fertility, recycling the vineyard's pomace from the vineyard's fruit. The deliberate creation of habitat and encouragement of biodiversity are important for the vineyard's general ecological well-being, as well as for providing a source of generalist predators and parasitoids for pest management. To do this, you can plant cover crops and create or conserve habitat in noncrop areas that support diverse beneficial creatures. Also important are the use of biodynamic preparations that influence the making of compost, soil health, and vine health. These compounds are applied in small amounts at key points in the season. Proponents describe these materials as "bioregulators"; their effects cannot be explained just on the basis of added nutrients or other easily measured parameters. Scientific evidence of their effectiveness is limited (Reeve et al. 2005). Many biodynamic winegrowers also integrate animal agriculture into their vineyards, grazing sheep and geese there during winter months and keeping poultry in the vineyard for meat and egg production (eggs are used to fine the wine) and for insect control (cutworms, earwigs, and other insects). When possible, biodynamic winegrowers collect the manure from their animal enterprises and incorporate it into the compost that they later apply to their vineyards. Finally, management decisions may also be based on the timing of cosmic cycles (solar, lunar, and sidereal time) for key practices in the vineyard and winery.

Biodynamics has somewhat of a "faith-based" approach that many laypeople and non-biodynamic agriculturists do not believe in or readily appreciate; hence, it is very controversial. And being somewhat based on faith, it is difficult or impossible to analyze or prove that some of the practices are effective. It is a management-intensive way to grow grapes and requires extra training and insight on the part of the grower to appreciate the subtleties and practices of the farming system. Many producers of high-quality wines use this system, a fact that continues to encourage interest, both among wine drinkers and among new growers considering a biodynamic approach.

Biodynamic vineyards are certified by a nonprofit group, Demeter USA. The USDA National Organic Program laws do not accept biodynamic certification as "organic," so many biodynamic vineyards are also certified as organic (when requested, Demeter USA does perform organic certification that is compliant with USDA NOP standards, as a separate service). In 2008, there were about 700 acres of biodynamic, Demeter-certified vineyards in California.

CONCLUSION

Organic winegrowing in California is still evolving and creative growers are still developing farming systems that fit their needs and are appropriate for the vineyards that they are managing. Organic winegrowers have demonstrated that they can farm in a way that is environmentally friendly, economically viable, and capable of achieving very high fruit and wine quality. Interest in these production practices continues to grow in California and around the world.

REFERENCES/RESOURCES

CDFA (California Department of Food and Agriculture). 2008. California organic statistics. www.cdfa.ca.gov/is/i_&_c/organic.html.

Koepf, H., R. Shouldice, and W. Goldstein. 1989. The biodynamic farm: Agriculture in the service of the Eatch and humanity. Hudson, New York: Anthroposophic Press.

Lotter, D. W., J. Granett, and A. D. Omer. 1999. Differences in grape phylloxera-related grapevine root damage in organically and conventionally managed vineyards in California. HortScience 34(6):1015–1047.

Reeve, J., L. Carpenter-Boggs, J. Reganold, A. York, G. McGourty, and L. McCloskey. 2005. Soil and wine quality in biodynamically and organically managed vineyards. American Journal of Enology and Viticulture 56(4):367–376.

Smith, R., K. Klonsky, P. Livingston, and R. DeMoura. 2004. Sample costs to produce organic wine grapes: Chardonnay: North Coast Region, Sonoma County. University of California Cooperative Extension GR-NC-04-0. http://coststudies.ucdavis.edu.

Weber, E., K. Klonsky, and R. DeMoura. 2005. Sample costs to produce organic wine grapes: Cabernet Sauvignon: North Coast Region, Napa County. University of California Cooperative Extension GR-NC-05-0. http://coststudies.ucdavis.edu.

Willer, H., and M. Yussefi. 2006. The world of organic agriculture: Statistics and emerging trends, 2006. Bonn Germany and Frick, Switzerland: International Federation of Organic Agriculture Movements (IFOAM) and Research Institute of Organic Agriculture (FiBL).

PHOTO: TOM LIDEN

A Brief History of the Ecological Farming Movement

GLENN T. McGOURTY

The ecological farming movement advocates for farming systems that are productive, economically viable, and focused on resource conservation. While there are different approaches and methodologies, they share certain common goals that include protecting air and water quality, maintaining fertile soils, encouraging healthy and diverse ecosystems (both on the farm and in surrounding areas), using productive, efficient, profitable farming techniques, encouraging family farming, and maintaining strong, vibrant rural economies and culture. This chapter presents a brief history of key players and events that have helped to shape this movement since the 1920s.

Following World War II, improvements in mechanization, fertilizers, and crop hybrids and the development of new pesticides collectively led to the greatest increase in farm productivity in U.S. history. Intensive animal husbandry increased the abundance of animal products and reduced their cost. Food processing increased, and many new convenience foods became available. By the end of the 1950s, refrigeration, efficient transportation, and marketing made the U.S. food system the envy of the world for its diverse, low-priced food, available year-round and offered in large, attractive supermarkets.

The intensification of agriculture accelerated long-term patterns of change in the rural landscape. Farming became more capital intensive, favoring larger farmers who could afford the high fixed costs of new machinery and facilities. The trend toward fewer farmers managing larger farms, which had been progressing for decades, sped up significantly. Many rural families who were unable or unwilling to follow this trend soon found themselves migrating from the farm to the city.

The farmers who remained altered the landscape to facilitate the use of ever-larger machinery. Many naturally vegetated areas, small riparian channels, and other topographic features were removed to make room for crop production. This loss of habitat decreased the biodiversity of animals and insects. Roadsides and drainage areas were sprayed with herbicides to eliminate all vegetation and to reduce sources of weeds and associated insects and diseases near fields and crops.

The widespread adoption of these new technologies inspired a new way of farming that changed agriculture across the globe. The term "Green Revolution" was introduced into the public discourse in 1968 by William Gaud of the U.S. Agency for International Development. Funded by the Rockefeller and Ford Foundations, Norman Borlaug led a group of researchers whose objective was to demonstrate that increased farm productivity could be brought to third-world countries using new crop cultivars that responded more efficiently to fertilizer, irrigation, pesticides, and mechanization. The idea was that by increasing farm productivity, third-world countries could eliminate chronic hunger. This proved to be substantially correct: in the modern era, sufficient food has been produced to feed a large and growing world population. Borlaug received a Nobel Peace Prize in 1970 for his work. In many respects, the Green Revolution is the model for modern conventional agriculture. However, the farming systems used by the Green Revolution were not always feasible in all settings due to their capital- and energy-intensive nature, and many subsistence farms in resource-limited regions remain in poverty.

By the 1950s, some unintended consequences of modern industrial farming began to appear. Intensive row cropping required tillage, which led to extreme erosion in some areas that were poorly suited to that practice. Soluble fertilizers and long-lived pesticides moved into surface- and groundwater, with the result that some rural lakes and streams became polluted from runoff containing phosphates and nitrates. Occasionally, fish populations were killed when pesticides drifted into waterways from large-scale spraying programs intended to control mosquitoes and forest insect pests (Carson 1962). Insecticide resistance began to develop, and resistant insects consequently required either higher application rates for existing pesticides or the development of new types of pesticides. Arthropod species that previously were not considered damaging, such as spider mites, became problematic when broad-spectrum insecticides harmed the many generalist predators and parasitoids that formerly kept them under control. Growers and crop professionals became convinced that crop production was no longer economically possible without pesticides. Some referred to these developments as a "pesticide treadmill." All of this ultimately encouraged the beginning of integrated pest management (IPM) practices.

Agrichemical dealers encouraged growers to make fertilizer and pesticide applications "by the calendar" without even determining whether a problem or need existed for them. After a while, some farmers began to notice a change in the health of their soil and crops. Next, overproduction became a problem, bringing down both crop prices and profit margins. Farming was becoming more and more capital intensive, making it less and less possible for a family farm to survive without someone in the family getting a job off the farm.

Then in 1962, a book by Rachel Carson forever changed agriculture in the United States. Carson's book, *Silent Spring,* detailed the loss of birds and damage to aquatic ecosystems from persistent pesticides, particularly the chlorinated hydrocarbons that were introduced after World War II and were now accumulating in animals in the food chain. America's national symbol, the bald eagle, was among the species threatened with extinction from chlorinated hydrocarbons, which weakened the shells of the nesting birds' eggs. Rachel Carson imagined and described the specter of cropland and its environs devoid of the sound of birds and wildlife. Despite technical criticisms by some in the commercial agriculture and pesticide industries about the accuracy of Carson's claims, *Silent Spring* became a best-seller and inspired many Americans to become advocates for wildlife and nature. Carson also awakened Americans' awareness of the interdependence of life, and how actions in one landscape affected life in other parts of the globe. Within 10 years of the publication of *Silent Spring,* a changing public sentiment resulted in the banning of the persistent chlorinated hydrocarbon pesticides, including DDT, heptachlor, dieldrin, chlordane, and others. Many consider the publication of *Silent Spring* as the beginning of the environmental movement in the United States.

In the 1960s, economic entomologists began developing more environmentally sensitive and effective crop protection methods. Scouting fields to determine pest numbers, targeting insects at the most susceptible stages of their life cycle, and using only the most effective insecticides were the first steps in a movement that was initially intended only to save money on pest management. The identification and use of beneficial insects were also advocated. The deliberate selection of cropping patterns and cultural practices to reduce plants' susceptibility to pests and diseases was also studied and encouraged by these new crop advisors. Like-minded professionals formed the Association of Applied Insect Ecologists and found themselves among the founding members of a new discipline called "integrated pest management" (IPM). IPM practices include those mentioned above and generally emphasize the use of preventive practices as a strategy to control pest species.

In the 1970s, IPM became more broadly adopted as the USDA began to fund IPM projects around the country. California started to license pest control advisers (PCAs) about the same time. PCAs who worked for farmers had an interest in using more precise pest management practices and materials that were effective, cost efficient, and less toxic to beneficial insects and to the people applying them. In 1979, the University of California established the Statewide

IPM Project as a way to advocate for and implement the IPM philosophy in pest management. IPM was viewed suspiciously at first by some agriculturists, but today it is an important part of production agriculture. IPM concepts have been widely adopted and are now used by most agricultural sectors, including organic and biodynamic producers.

A severe downturn in the nationwide agricultural economy in the 1980s forced agriculture to once again re-evaluate its direction. At this point the concept of sustainable agriculture was developed as an alternative approach to conventional agriculture, which was still so dependent on petrochemical inputs and pesticides. These inputs became increasingly unaffordable as farm prices fell and profit margins shrank.

Advocates and visionaries saw sustainable agriculture as a way for growers to reduce their costs of production and at the same time reduce adverse environmental impacts on their farms. They strived to take advantage of natural processes and beneficial on-farm biological interactions, reduce off-farm inputs of energy and nutrients, and improve the efficiency of their operations to increase profitability. A sustainable agriculture practitioner understands his or her farm as an ecological system in which the many interactions of soil, climate, crop growth, and pest management are studied and then optimized to produce crops.

The stated objectives of the sustainable agriculture movement are to sustain and enhance rather than reduce and simplify the biological interactions on which production agriculture depends, thereby reducing the harmful off-farm effects of production practices. Besides profitability, sustainable agriculture also takes into account the environmental effects and social consequences of farming practices and agricultural policies.

Sustainable agriculture encompasses many approaches to farming, including organic and biodynamic agriculture as well as farming livestock with intensive management and integration into plant production (intensive grazing, rotating pastures, forage crops, and row crops grazing in orchards and vineyards), reduced use of pesticides and herbicides, profitable and efficient production with an emphasis on improved farm management and conservation of soil, water, energy, and biological resources, exploration of new crops and alternative markets, and the preservation of family farms and rural communities.

Farms have come to be viewed as ecological systems in which the many interactions of soil, climate, crop growth and pest management should be recognized and used to manage crops and associated pests and diseases. In theory, manipulation of these beneficial biological interactions allows farmers to reduce off-farm inputs of nutrients and pesticides and improve the profitability of their operations. This is sometimes referred to as a systems approach (D. Pimentel and M. Pimentel in Knorr 1983; Kirschenmann 1988).

Sustainable agricultural farming systems are designed to be environmentally sound, economically viable, and socially just over time. While the legislation that defines organic farming does not specify social and environmental stewardship objectives, many organic growers are well versed in issues specific to sustainable agricultural production and actively support efforts to aid farmworkers and rural communities, preserve agricultural land, and promote other activities that protect the environment, wildlife, and biodiversity (A. Thrupp, personal communication). By the late 1980s, a large number of papers and books had appeared discussing agricultural sustainability, and most professional societies concerned with agriculture had adopted statements about sustainability as guidelines (Edwards et al. 1990). Many agriculture schools in the United States began to offer classes and coursework on the practices and policies of sustainable agriculture.

The sustainable agriculture movement also promotes different marketing approaches that are more profitable for otherwise marginalized growers, such as farmers markets, Community-Supported Agriculture (CSAs), farm trails, and other ways to help small producers sell directly to consumers for maximum profitability.

THE BIODYNAMIC FARMING MOVEMENT

The first clearly defined alternative farming movement in the twentieth century was biodynamic farming. Its originator, Rudolph Steiner, was born in 1861 in what is now Croatia. Steiner grew up in the midst of a modernizing society superimposed on a traditional peasant culture whose way of life had changed little for many centuries. He was considered a philosopher and critic and was the editor of a successful cultural magazine. He founded the Anthroposophy Movement to explore and share his interests in spiritual research based on science. He also founded the Waldorf School system and taught about homeopathic medicine.

In 1924, Steiner presented a series of eight lectures to a group of farmers and others in Silesia (Germany) who were concerned about the decline in the health and quality of their farms, livestock, and crops. Steiner took a holistic view of agriculture and regarded the farm as an organism, a concept that is in some ways analogous to current discussions of the farm as a system. He reasoned that since plant growth is dependent on the sun, earth, air, and water, the entire universe affects the processes of life. Steiner believed that life exists in a nexus between an ethereal world of atmosphere, light, and cosmic energy and an inanimate world of darkness, earth, and stone (Koepf, Shouldice, and Goldstein 1989).

In his lectures, Steiner explained the need to organize farms in a way that addresses the basic needs for fertility, and pest management, as a self-sufficient system. He emphasized creation of a healthy environment for crops, farm families, and the communities that they live in, while still maintaining an economically viable agricultural business. Much more than was common for the conventional agriculture of his time, Steiner promoted nutrient conservation and recycling practices and the careful integration of ruminant livestock such as dairy cows with crop growing as essential for creating the balances needed to make a biodynamic farm successful. He did not discuss the production of horticultural crops or market gardens. Modern biodynamic farm operators follow many sound agronomic principles, including crop rotation, integration of livestock and crop production, and careful composting. They also avoid the use of synthetic fertilizers and pesticides.

While most agriculturists recognize two solar cycles (daily and yearly), Steiner concluded that three lunar cycles, sidereal time (that is, time related to the movements of the stars and planets), and other natural rhythms also affect animal and plant growth. Biodynamic farmers believe that the benefits of certain agricultural practices are enhanced or diminished by varying their timing in relation to celestial influences, so they plan their farming operations according to the celestial calendar as well as the solar calendar. In his lectures, Steiner also described the use of a set of "preparations" that are extremely controversial among conventional agricultural scientists. Steiner's contention was that these biodynamic preparations influence life processes in plants, compost, and soils. He said that they help growing plants make appropriate use of the many forces in their environment, such as the living soil, neighboring plants, the atmosphere, sunshine, and etheric and astral forces. There are eight materials (numbered 500 to 507) that are specially prepared and then applied in diluted form at homeopathic levels (fractions of an ounce to the acre) to either plants, soil, manures, or compost, at specific times of the year. The action of these preparations is hard to quantify because they are said to function as bioregulators. Their effects cannot be explained solely in terms of response to their mineral or nutrient content; rather, they are said to influence esoteric properties.

There is now a large and varied scientific literature documenting tests of these preparations and their effect on food quality, much of it published in German (Koepf, Shouldice, and Goldstein 1989). Researchers' results are inconsistent. Some studies have shown that in many cases the economics of biodynamic farming are favorable in comparison to conventional agriculture and that growers can be successful using this very unconventional farming system (Koepf, Shouldice, and Goldstein 1989). Studies have documented that soil quality and fertility can be very good on biodynamic farms (Reganold 1988). Most biodynamic farms are in Europe, but there are examples in many locations throughout the world, including California.

ORGANIC AGRICULTURE

Biodynamic agriculture is a form of organic agriculture, but the organic agriculture movement, which is more widely known in the United States, originated in Britain in the early twentieth century. Its most well-known figure is Sir Albert Howard, a British agriculturist who was educated at London and Cambridge Universities in the late 1800s.

Howard worked as an economic botanist at the Agricultural Research Institute at Pusa in northeast India. His assignment was to create improved crop cultivars, but he soon concluded that it would be futile to try to breed new materials unless an effort was also made to gain an understanding of the tropical Indian farming system and its conditions. His work led him to conclude that the good health of agricultural crops and animals was based on fertile and healthy soils. He argued that the humus (carbon) content of the soil was essential for overall soil fertility. This countered the prevailing belief of the times that a soil's fertility was determined mostly by its inorganic mineral content.

Beginning in the 1840s, German chemist Justus von Liebig discovered procedures for determining the mineral content of crops by reducing the crops to ashes and then analyzing the remains for major minerals. He showed that by adding these elements as mineral salts, one could solve soil fertility problems. This discovery led to the development of modern soil chemistry and to the practice of fertility management based on the use of mineral fertilizers.

Liebig's approach to plant nutrition was based strictly on inorganic chemistry and ignored biological interactions. In contrast, Howard argued that soil fertility programs that were based solely on mineral nutrition would fail over time as a result of erosion and loss of soil structure. Working in tropical India, he found that when growers used only chemical fertilizers, micronutrient deficiencies became a problem in many tropical soils, which tended to have limited mineral reserves. The fertility of these soils was held in the organic fraction (the humus) as much or more than the inorganic fractions (the sand, silt, and clay). He recognized the roles of soil microorganisms, earthworms, and different root morphologies of crops, and how the

ecology of a farm could be used to manage resources for maximum effectiveness. Howard could be considered one of the first agroecologists because of his interest in all of the organisms that make up a farming system, ranging from microbes to annual crops, trees, and farm animals. This systems approach to agricultural research was very forward thinking and is only now starting to be more widely used by agriculturists and researchers.

Howard also studied how nutrients were cycled through this system and how changes in the system caused predation and disease. His idea that "health is the birthright of all crops" was thought of as rather radical at the time. In 1931, he published the book *The Waste Products of Agriculture: Their Utilization as Humus*, which detailed his studies of composting. He realized that manure alone could not supply all of the nutrients required by farms and that crop residues could greatly extend the nutrients contained in manures and replenish carbon in the soil. A second book published in 1940, *An Agricultural Testament*, summarizes his observations during 25 years of research in India. In many respects, it was the first publication on an organic approach to farming written in modern times. The foundations of agroecology, agricultural systems research, and organic agriculture can all be found in his writings and efforts. Howard and others developed ideas about the connections between human nutrition, soil quality, and pest and disease resistance that still find advocates in the organic farming movement today.

In the late 1930s, J. I. Rodale bought property in rural Pennsylvania in order to pursue his interest in farming. He had read the works of Howard and other alternative agriculturists. In 1947, Rodale founded the Soil and Health Foundation to help showcase practical methods for rebuilding natural soil fertility and to promote the benefits of healthy diets. His activities included demonstrations and applied research. He funded his activities with profits from a modestly successful publishing house that produced the magazines *Prevention*, *Organic Gardening and Farming*, and the *Health Bulletin*. He subscribed to Howard's ideas that our health is tied to our food supply and that food quality is tied to the management of soil fertility through the organic fraction of the soil.

By the time of his death in 1971, J. I. Rodale had built a loyal following and the organic movement was rapidly becoming established in American culture. At that time, most organic farming efforts were confined to backyard gardening and small-scale agriculture. As public interest in organic products grew, though, so did the demand. Growers struggled at first to develop organic farming techniques. Outside of the biodynamic movement, there were few sources of professional advice for larger-scale organic farming operations. Over time, this has changed, and now California is the leading U.S. state for the production and export of organic products.

In 1973, the organization of California Certified Organic Farmers (CCOF) was founded to help monitor and legitimize organic farmers. It also became an important political organization, represented organic farmers as a trade association, and was the first such organization in the United States. CCOF helped to legislate the California Organic Food Act in 1979, formally defining organic farming from a legal perspective. In 1990, California passed the California Organic Foods Act, strengthening the previous legislation to include legal sanctions administered by the California Department of Food and Agriculture. The legislation also required registration of organic growers with county agriculture commissioners, a three-year transition period for conversion of conventionally farmed cropland to organically certified ground, and a listing of accepted materials for fertilizing and pest management.

In 1990, the U.S. Federal Farm Act established national standards for use of the term "organic." The guidelines generally followed those set forth in the California Organic Foods Act. The USDA undertook a National Organic Program (NOP) during the 1990s to standardize organic regulations for all states. Rules were first published in 1998, and after extensive public comment they were rewritten and the NOP implemented in 2002.

Organic farming has gone from a marginal business with a small following to a rapidly increasing sector of the food system. By the 1990s, sales of organic products were enjoying double-digit growth nearly every year. Supermarket-sized stores based on organic products were established and have

since become commercially very successful. Large corporations began growing, processing, and selling organic products. Even conventional grocery stores began to offer organic products. Stores that managed their organic food sections with trained, knowledgeable staff in supportive (usually affluent) communities were profitable. The NOP helped to standardize labeling and certification and has helped to build a strong, consistent sense of the meaning of the term "organic" among consumers. The NOP also created the structure for enforcing the regulations and code throughout the United States.

SUSTAINABLE AND ORGANIC WINEGROWING

Organic winegrowing began in the 1970s when idealistic growers wanted to raise high-quality fruit without using synthetic chemicals. Most organic producers were small scale. They developed a farming system using techniques taught by "old-fashioned" growers, such as hoe plowing beneath vines, use of legume cover crops to add nitrogen to the soil, and use of sulfur and copper dusts for disease control. Their system worked surprisingly well: the early organic winegrowers were able to grow normal-sized crops of high-quality fruit and were making economic returns similar to those enjoyed by conventional growers. In the early 1990s, Fetzer Vineyards in Mendocino County became one of the first large-scale farming interests to certify a large acreage—over 1,400 acres of vines—as organic. Today California has nearly 11,000 acres of organically certified vineyards, with 4,000 acres just in Lake and Mendocino Counties. Many more acres are farmed organically but have not been certified. The term "organic winegrowing" is well defined in law and custom and can only be used by certified growers.

Another effort, biodynamic winegrowing, began in California as the results of the early 1990s efforts of the Fetzer family and Fetzer Valley Oaks gardener Michael Maltas, who studied biodynamics at Emerson College in England under noted professor Herbert Koepf. With the assistance of biodynamic consultant Alan York, Jim Fetzer became Demeter certified in biodynamics at his Ceago winegarden and at the family's Redwood Valley property in the mid-1990s. Other Fetzer family members have certified their properties as well. The interest in biodynamic winegrowing continues to grow, and there are now 70 Demeter-certified properties in Mendocino, Lake, Napa, Sonoma, and Monterey Counties. Biodynamic practitioners believe that their techniques can allow them to exceed the quality standards of organic winegrowing. Biodynamic winegrowing is also widely practiced in Europe, especially in France and Germany, where more than 400 biodynamic producers are in operation, many of them producing very high quality (and expensive) wines.

The Code of Sustainable Winegrowing Practices adopted as a joint effort by the California Association of Wine Growers (CAWG) and the Wine Institute has set higher environmental standards for the wine industry. This voluntary self-assessment program has been widely implemented in many regions of California. According to the code's definition, "sustainable winegrowing" means that the grower and winemaker follow growing and winemaking practices that are sensitive to the environment, responsive to the needs and interests of society at large, and economically feasible to implement and maintain. Growers are made aware of farming practices that have a smaller impact on the environment while still maintaining productivity and good economic returns. Under the code, wineries are encouraged to reduce their energy use, reduce and recycle their waste stream, and minimize the use of hazardous materials. Both growers and wineries are encouraged to attract and retain excellent employees through training, incentives, and fair labor practices. Finally, the code encourages techniques that create good relations with neighbors and the larger community in an effort to promote vibrant rural communities that are economically viable and pleasant to inhabit.

Certification programs are now available for sustainable winegrowing, including Lodi Rules (administered by the Lodi-Woodbridge Winegrape Commission) and others.

CONCLUSION

Winegrowers in California continue to look for new ways to grow high-quality fruit and make flavorful wines in economically viable, environmentally sound, and socially responsible ways. Alternative farming practices based on the principles of sustainable agriculture help winegrowers to protect resources and stay in business and at the same time provide wine drinkers around the globe with a wide range of flavorful, high-quality wines produced in environmentally friendly ways. By adopting these practices, growers may also open new market niches for their products, especially among wine drinkers who support environmentally friendly farming and food processing practices.

REFERENCES/RESOURCES

Carson, R. 1962. Silent spring. Boston: Houghton Mifflin.

Dlott, J., C. Ohmart, J. Garn, K. Birdseye, and K. Ross. 2002. The code of sustainable winegrowing workbook, 1st ed. San Francisco: Wine Institute and the California Association of Winegrape Growers.

Edwards, C., R. Lal, P. Madden, R. H. Miller, and G. House. 1990. Sustainable agricultural systems. Ankeny, Iowa: Soil and Water Conservation Society.

Kirschenmann, F. 1988. Switching to a sustainable system: Strategies for converting from conventional/chemical to sustainable/organic farming systems. Lamour, North Dakota: Northern Plains Sustainable Agricultural Society.

Knorr, D., ed. 1983. Sustainable food systems. Westport, Connecticut: AVI Publishing.

Koepf, H., R. Shouldice, and W. Goldstein. 1989. The biodynamic farm: Agriculture in the service of the Earth and humanity. Hudson, New York: Anthroposophic Press.

National Research Council (U.S.). 1989. Alternative agriculture. Washington, D.C.: National Academy Press.

Reganold, J. 1988. Comparison of soil properties as influenced by organic and conventional farming systems. American Journal of Alternative Agriculture 3:133–155.

Organic Certification and Registration in California

DAVID CHANEY, RAY GREEN, AND L. ANN THRUPP

BACKGROUND

The organic industry grew significantly during the 1990s and early 2000s. For the United States as a whole, the total market value (retail sales) of organic food products, including processed products, grew from about $1 billion in 1990 to an estimated $7.8 billion in 2000 (Dimitri and Greene 2002), while the Organic Trade Association estimated 2006 organic food sales at around $16.7 billion (OTA 2007). California producers have led this trend, showing an increase in both numbers of organic farmers and total acreage. From 1992 to 2003, the number of registered organic farms in California grew by almost 30 percent, from 1,273 to 1,765 growers. Over the same period organic acreage quadrupled, increasing from 42,000 acres in 1992 to almost 172,000 acres in 2003 (Klonsky and Tourte 1998, 2002; Klonsky and Richter 2005).

Prior to 1990, few common labeling standards were agreed upon and for the most part there was no regulatory requirement for organic certification. As the market began expanding, organic producers and marketers recognized the need for more uniformity and integrity in their products and they turned to Congress for assistance in developing national standards to govern organics. The result of these efforts was the Organic Foods Production Act (OFPA) passed in 1990 as part of the Food, Agriculture, Conservation, and Trade Act. OFPA mandated that the U.S. Department of Agriculture (USDA) establish an organic certification program for producers and handlers of agricultural products who sell and label their products as being organic. Responsibility for developing and administering a National Organic Program (NOP) was assigned to USDA's Agricultural Marketing Service (AMS). The intent of OFPA was to establish national practice standards governing the production and marketing of certain organically produced agricultural products, to assure consumers that organically labeled foods represented these consistent standards, and to facilitate interstate commerce in fresh and processed food that is organically produced and labeled as such.

California did have an organic labeling law in place starting in the 1970s, but the law had no enforcement component. Concurrent with the development of OFPA, California passed its own law regulating organic farming, the California Organic Foods Act (COFA), which was signed into law in 1990. COFA set a voluntary certification standard with full registration and enforcement components. The California law became effective immediately upon signing, but it took more than 10 years for the federal OFPA to be completed and implemented. When COFA was eventually superseded by the federal regulations, the law was rewritten to become the California Organic Products Act of 2003 (COPA).

The process of completing the national rule was complicated and controversial, involving numerous drafts and public input from consumers and the organic industry. USDA issued a proposed rule in 1997, but the final program was not completed until December 2000. The final rule states that all organic producers who gross more than $5,000 per year in retail sales must be certified through an accredited certification agency. These certifying organizations, both public and private, act as third-party agents of the NOP to legally verify that production and handling practices meet the national standards. Enforcement of the NOP began in October 2002, and AMS immediately began to accredit certification agencies to assist with its implementation.

In addition to accreditation of certifiers, the NOP is also responsible for ensuring that the purposes of the OFPA are accomplished, determining the equivalency of foreign programs for imports into the United States, participating in the development of international standards, coordinating enforcement activities with other agencies, conducting the petition process for materials review, and providing administrative support for the National Organic Standards Board (NOSB). The NOSB is a 15-member committee established as part of OFPA to help the Secretary of Agriculture develop standards for substances to be used in organic production and to advise the Secretary on other aspects of implementing the program. The current board includes four farmers/growers, two handlers/processors, one retailer, one scientist, three consumer/public interest advocates, three environmentalists, and one certifying agent.

STATES AND THE NOP

State governments can serve two functions in relation to the NOP: they can either choose to implement a State Organic Program (SOP) or become an accredited certifying agent, or they can choose to do both. An SOP is primarily an enforcement program and is not involved in certifying growers or handlers. Once approved as an SOP, the agency is responsible for all aspects of enforcement within the state. A state agency that becomes an accredited certifying agent serves the same function—certifying organic production, processing, and handling—as any other certifying agency would. In the list below, you can see how three states are working in this area.

- **California.** The California Department of Food and Agriculture (CDFA) administers an SOP only and does *not* certify.

- **Washington.** The Washington State Department of Agriculture is an accredited certification agency only, not an SOP.

- **Utah.** The Utah Department of Agriculture and Food administers an SOP and is also an accredited certifying agent.

In order to qualify as an SOP, a state must adopt its own set of regulations or laws that accept the standards set by the NOP. Upon approval by USDA, those state laws essentially become the NOP laws for that state. COPA (the current California law), for example, incorporates by reference the federal regulations and is interpreted in conjunction with them, and it is written so that any future changes to the federal law automatically become regulations for California. Compliance and enforcement provisions are part of both the NOP and approved SOPs. Most states do not have SOPs; they fall under the direct jurisdiction of the federal NOP with regard to enforcement activities.

NOP IMPLEMENTATION IN CALIFORNIA

California's SOP is administered through CDFA (responsible for fresh agricultural commodities [e.g., fruits, nuts, vegetables, field crops], meat, poultry, and dairy products) and the California Department of Public Health (CDPH) (responsible for processed food products, cosmetics, and pet food). The SOP's purpose is to protect producers, handlers, processors, retailers, and consumers of organic foods sold in California by enforcing labeling laws related to organic claims for agricultural products. Its activities are coordinated with the California Organic Products Advisory Committee, USDA, and the California county agricultural commissioners.

The California SOP's responsibilities include administration of the program, training county staff, investigating complaints, registering producers and certification organizations, and acting as a resource for information on COPA and the organic industry in California. The SOP is funded entirely by producer and certifier registration fees, a portion of which are used to support county enforcement activities. Because COPA brings two separate statutes relating to fresh and processed products together into one law, it is important that each individual carefully read the sections that relate to his or her type of business operation. The complete text of COPA is available online at the CDFA Organic Program website, www.cdfa.ca.gov/is/i_&_c/organic.html. If you are involved in both aspects—growing and processing—you should note the specific requirements pertaining to each.

Producers

Organic producers and handlers (e.g., wholesale distributors, retailers, and some processors who market direct to retail or consumers) are regulated through CDFA. California requires that organic producers be registered and, depending on their sales revenue, certified as well. All organic growers, regardless of gross sales, must be registered with CDFA. In addition, any producer who grosses more than $5,000 per year in total sales of organic products must be certified through an accredited certification agency.

Registration, which involves providing a map of the production area, a list of crops that you intend to produce as organic, and a 3-year history of substances or materials applied, is handled by your local county agricultural commissioner. You have to provide verification of the land use history and pay an initial minimum registration fee of $75, or a higher fee if gross sales are or are expected to be higher than $25,000 annually (see Table 3.1). After initial registration, fees for yearly renewal follow the schedule in table 3.1. The actual total registration fee follows a schedule based on your annual gross organic sales. Fees for yearly renewal of registration follow the same schedule.

Certification follows a separate process. NOP regulations establish who must become certified before they can use the "organic" label on their products. The following general guidelines apply:

- **Exempt from having to be certified: Producers who are grossing less than $5,000 per year in retail sales.** Product must be produced, processed, and packaged under the producer's own label; product cannot be sold to anyone else for processing or packaging, although under the

producer's own label the product can be marketed in retail stores.

- **Excluded from requirement of having to be certified: Any operation or company that does not repackage or relabel the product** (e.g., bulk wholesale distributor) or retail establishments that process or prepare organic product on the same premises at which they sell to the customer.

- **Must be certified: All other operations involved in production, processing, and marketing.**

If you are not sure whether you need to be certified or what category you fall into, contact the CDFA SOP office at (916) 445-2180 (website: www.cdfa.ca.gov/is/i_&_c/organic.html). The steps involved in certification are described in more detail under "Certification Agents" later in this chapter.

Processors

As defined in COPA, every person in California who processes, packages, stores, distributes, or handles food, pet food, or cosmetics in California that are sold as organic (with the exception of processed meat, fowl, or dairy products) is required to register with the California Department of Public Health. Registration is handled through CDPH's Organic Processed Product Registration Program, and processors must pay an annual registration fee. Like producers, they must be both registered and certified.

CDPH itself handles registration of organic processors. To get an application, call CDPH or visit the CDPH website, www.dhs.ca.gov/fdb/html/food/organicreg.htm. The application must be filled out completely and returned to CDPH with the appropriate fee. Upon receipt of the application and fee, the operation is registered, provided that if the operation processes food for human consumption it already has a valid processed food registration permit issued by CDPH. There is no pre-inspection or verification requirement, but during the operation's regular food safety inspection, the inspector may ask questions about the organic aspects of the operation. The registration fee is set according to the schedule shown in Table 3.2. Fees for yearly renewal follow the same schedule.

It is possible that an operation or business will have to register both as producer and processor with both agencies (CDFA and CDPH). For example, olive growers who also make and sell olive oil must be registered with both CDFA and CDPH. The

Table 3.1. Fee schedule for companies that are required to be registered with CDFA*

Gross annual sales	Registration fee
$0 – 4,999	$25
$5,000 – 10,000	$50
$10,001 – 25,000	$75
$25,001 – 50,000	$100
$50,001 – 100,000	$175
$100,001 – 250,000	$300
$250,001 – 500,000	$450
$500,001 – 1,000,000	$750
$1,000,001 – 2,500,000	$1,000
$2,500,001 – 5,000,000	$1,500
$5,000,001 – 15,000,000	$2,000
$15,000,001 – 25,000,000	$2,500
$25,000,001 and above	$3,000

*Fees are subject to change: Check with CDFA for the current fee schedule. Producers who sell processed product pay fees based on the value of the raw product prior to processing and the value of any product sold as unprocessed. Other exceptions are outlined in the text of the California Organic Products Act, available online at http://www.cdfa.ca.gov/is/i_&_c/organic.html.

registration fee for CDFA is based on the market value of the olives prior to processing, whereas the fee for CDPH is based on the market value (gross sales) of the finished olive oil.

Certification, which is distinct from registration, involves the same steps and requirements for processors as it does for growers (see the section "Certification Agents" later in this chapter). The same restrictions and exemptions described above for producers also apply to processors. If your farming business is involved in both production and processing, it makes sense for you to find a certifying agency that is accredited to certify both aspects (production and processing) of the operation. Most certifiers do both, but it is wise to make sure. Some certifiers consider postharvest handling to be a separate activity requiring a separate application, a fee-for-service charge, and inspections. Other certifiers see processing as an addendum or attachment to the production side of the operation, an approach that can reduce costs and paperwork for the grower. This discretion on the part of the certifier only goes just so far: any activities beyond simple postharvest handling (e.g., beyond storage or transportation to market) would clearly fall under the "processing" rules and require separate registration and certification for each component.

If you are not sure whether you need to be certified or what category applies to your operation, contact the CDPH Food Safety Program at (916) 650-6500. The remainder of this chapter covers the certification process and is written mainly for growers, but the general principles and steps involved in becoming certified apply to processors as well.

Table 3.2. Fee schedule for companies that are required to be registered with the California Department of Public Health (CDPH)*

Gross annual sales or revenue	Annual registration fee
$0 – 5,000	$50
$5,001 – 50,000	$100
$50,001 – 125,000	$200
$125,001 – 250,000	$300
$250,001 – 500,000	$400
$500,001 – 1,500,000	$500
$1,500,001 – 2,500,000	$600
$2,500,001 and above	$700

*Fees are subject to change: Check with CDPH or CDFA for the current fee schedules.

Certification Agents

An organic producer can choose from any of the certification agencies accredited by USDA, and should select the one that will best serve his or her needs and budget (see Table 3.3 and Sidebar 3-A). As of spring 2008, more than 90 accredited certification agents had been approved by the USDA, including state and international agencies as well as private companies. About 20 of those were registered under the California SOP. All certifiers who want to operate in California must register with CDFA. Most certifiers require an application fee as well as inspection fees. These are substantial, starting at around $500. If you intend to export your operation's organic products outside of the United States, you will also need certification from the certifying body in the destination country or from the International Federation of Organic Agricultural Movements (IFOAM). Some certifiers that are accredited through the NOP have direct partnerships with those foreign certifiers, meaning that they are accredited by those international agents and can verify international certification in addition to NOP certification.

The certifying agency monitors grower practices to ensure that the grower is in compliance with NOP regulations. The certifier

- must know and enforce the NOP standards

- must verify the producer's compliance with national standards through annual scheduled inspections of the client's records, fields, and production and handling areas

- may perform annual soil and tissue tests from the client's operation in accordance with NOP guidelines

- may perform a certain number of surprise inspections every year

- must conduct ongoing review and inspection of client operations to ensure compliance

As a third-party agent of the NOP, the certifier is forbidden to provide advice on the producer's operation but can and must provide information to the producer about the certification process and the legal requirements for maintaining certification. This means that the certifier should know the client's operation in detail and should have current knowledge of NOP regulations and be able to communicate

Sidebar 3-A

HOW TO EVALUATE AND SELECT A CERTIFIER

Questions to ask yourself before you begin contacting certifying agents:

1. Will my product be sold outside of California? If not, you may want to contact only those agents with home offices in California.

2. Will my product be sold outside the United States? If not, you only need NOP certification. If you are marketing overseas, you need to look for agents who certify according to international standards (e.g., Japanese Agricultural Standard [JAS], European Union [EU], International Federation of Organic Agriculture Movements [IFOAM]) and that have knowledge of those markets.

3. Is my "product" a raw crop or a processed food? Which agents seem to be strongest and provide the best service in relation to my final product?

4. Who certifies my friends, competitors, and other companies like mine in the county or state? Talk to those companies and find out how satisfied they are with their certifying agent.

The NOP's official *List of Accredited Certifying Agents* is published on the NOP website at www.ams.usda. gov/AMSv1.0/nop. The *New Farm Guide to US Organic Certifiers* is a very useful tool for learning about and comparing accredited certifying agents in the United States. For more information, visit the Rodale Institute website at www.rodaleinstitute.org/certifier_directory.

Questions to ask certifying agents after you have checked the USDA NOP website and confirmed that the certifying agent is accredited by NOP:

1. Given the products/land/processing/etc. that I propose to certify, what are expected costs for the initial submission of a certification request, the expected costs of actual certification on the first inspection, and the expected ongoing annual costs? When are the fees due and payable?

2. Is your cost/fee structure linked to organic sales, overall sales, number of acres, or complexity of the operation?

3. Once paid, are any of my fees refundable if I decide to withdraw from application or from certification? Are there any additional costs or hidden fees?

4. Are there any discounts for renewal clients?

5. How many pages would a typical application package be for an operation similar to mine?

6. I am going to export to *(name of country or markets)*. What is your expertise in certifying for this target market? What documents or transaction forms are required for export of organic product to this country? Are there any fees for generation of specific documentation required for each individual shipment?

7. Will you sign and comply with a "nondisclosure" agreement?

8. What are the charges if I need to add or change something during the year (land, new crop, new product, new SKU of same product, etc.), and what documentation is required?

9. Do you offer training programs in any aspect of organic compliance?

10. Will you assist in any way in marketing organic commodities?

11. Do you have a website listing the companies you have certified? Do you list companies you have certified by specific commodity?

12. Will you certify co-handlers under my application? In that situation, would the fee be reduced?

13. How long have you been certifying organic operations?

14. What type of verification is needed to prove land history?

15. How do you determine allowed and prohibited materials? Other considerations:

All certifiers are private companies or are part of a governmental entity. Each will have a different ability to respond to you and to provide services. You can gauge the level of service by checking:

1. What is the average time that it takes the certification agency to complete a certification?

2. What is the average time that it takes the agent to call back or respond to a question that I may have?

3. Does the certifier provide supporting information that helps me understand compliance issues?

4. What kind of support does the certifier offer for helping me understand and complete paperwork?

5. Does the certifier provide a detailed explanation outlining the sections of the NOP that back up their certification positions and decisions?

6. Are the inspectors friendly and courteous? Are the inspectors part of the certifying company or does the company contract with independent inspectors? Does the producer have the option of choosing a particular inspector?

them effectively but may not act as a consultant to the grower. Certification agents must maintain their accreditation through the NOP and, as noted in the Table 3.3, may also become certification agents for foreign or international bodies. Some certifiers may also serve a broader role in organic trade associations or in dealing with policy issues.

The Organic System Plan and Growers' Responsibilities

Growers should recognize that they also have serious responsibilities and commitments in relation to certification. According to NOP regulations, every certified organic producer is required to develop and keep current a production or handling organic system plan (OSP) that describes the operation and the practices and procedures to be maintained and also provides a list of each substance to be used as a production or handling input. The OSP includes

- land history

- sources of all seeds and planting stocks used by the organic operation

- all inputs to the crop and all materials applied to the crop

- list of all practices and procedures for soil and pest management, fertility, and crop nutrition that are used in the operation, including details on monitoring practices

- explanation of barriers or buffers used to prevent commingling between organic and nonorganic products or drift from nonorganic operations

- harvest and postharvest practices, including equipment use

- a description of records kept to prove compliance

- any additional information needed to document NOP compliance

Most certifiers provide forms that growers can use to prepare their OSP. These forms may be helpful for some growers, but they are not required; it is possible to write your own OSP, so long as you include all the required information. In either case, writing the OSP is useful for planning your farming operations for the year. Some certifiers may provide help in preparation of the OSP. The length of time to complete an OSP depends on the level of complexity of the farming operation.

The OSP is central to the certification process. It serves as a management tool to help farmers make decisions and react to changing circumstances. It also describes the human and natural resources of a farm, helps a producer manage those resources in an integrated way, and can help the grower with budgeting and financial planning. Last and most important, the OSP constitutes a legally binding contract between the certifier and the certified operation. Breach of that contract can result in denial or loss of certification. This last point is crucial, and growers must understand that the records they keep constitute the only proof that their "contract" (the OSP) has been fulfilled. The OSP is the commitment or promise to the certifier that production and handling will be carried out in a certain way, but the grower must show through good record keeping how he or she has kept that promise. Growers must get approval from their certifier if they are going to deviate from their submitted organic system plan.

In summary, the federal OFPA is a performance regulation, not a prescriptive regulation. This means producers decide how they will grow or process their products and how they will demonstrate compliance with the NOP. The OSP is the individual grower's or operator's specific plan on how to meet those performance targets. Growers should be proactive and create the road map for achieving compliance that works best for them.

THE TRANSITION AND STEPS FOR ORGANIC CERTIFICATION

The Transition Period

In order to be certified organic, a farming site cannot have had any prohibited substances applied to it for 36 months. One way to accomplish this is to choose a site that has not been farmed for at least the last 3 years. If the site has been actively farmed, the grower must ensure (and verify with accurate records) that no synthetic, noncompliant materials were used during the 3-year period. The 3-year transition period is a time when growers can educate themselves about the NOP and certification requirements (see Sidebar 3-B and the list of Internet resources at the end of this chapter). This is also an opportunity to begin work on a soil-building program, an integral component of the OSP. A grower in transition may also consider hiring a consultant to help prepare the

Sidebar 3-B

TRANSITION ISSUES FOR ORGANIC WINEGROWERS

The switch to organic production involves major changes in production, management, and marketing. Before starting the transition, it's helpful for growers to consider the following factors and questions:

- Are you willing to make some significant changes in how you approach farming practices?

- Are you willing to spend more time on your production and business activities?

- Do you enjoy walking your fields on a regular basis?

- Are you enthusiastic about educating yourself?

- Do you have a buyer who is interested in organic grapes?

- Have you confirmed that there are market opportunities?

- Are you willing to develop new management skills and to accept some possible risks during the transition stage?

- Are there local organic growers, consultants, and other local resources available in case you confront challenges or questions?

One way to look at the transition period is to see it as an investment in educating yourself about organic farming and marketing. You can enhance the value and return on that investment of time by learning as much as possible about the ecology of farming systems and gaining new skills in managing your business and marketing your product. Here are some important goals to work toward during the transition:

- Identify the certification organization that is appropriate for you and collect information on the requirements and steps for that certification organization.

- Make contacts with local growers and extension agents and/or farm advisors who have experience with organic production.

- Gather information by attending educational seminars or workshops about organic practices and research relevant publications, written materials, and information from the Internet.

- Identify marketing strategies or options for your products and explore new relationships with potential buyers.

- Evaluate your current resources and practices and determine which of your current conditions and practices will need to change in organic farming.

- Identify the strengths and weaknesses in your vineyard, in terms of the natural resources, the pests and naturally occurring beneficial organisms, the

types of soils, vine vigor, climate and water availability, etc. A soil analysis is particularly essential so that you will be able to start with well-adapted strategies to build healthy soil under your particular local conditions.

- Focus on preventing problems and addressing the causes of problems using a systems approach. The switch to organic is not simply a matter of replacing conventional inputs with organic inputs. Learning about the systems approach beforehand is therefore important.

- Evaluate your weed management practices. Weeds pose a major challenge in the switch to organic production. Growers can spend time during the transition thinking about how they perceive and manage weeds. Many organic growers find that they have to make an attitudinal shift toward greater tolerance of weeds in mature vineyards and they come to see that some plants they would ordinarily see as weeds actually play a useful role in adding organic matter to soils. Organic production requires a balanced approach to vegetation management in the vineyard. Chapter 7 of this book provides more details of various weed management strategies.

- Start to work on your organic system plan (OSP) by identifying your goals and the "road map" of practices and steps that you will take to make the organic transition and to achieve your goals. Write your plan up so that it is clear. Your specific steps or processes may change slightly as you learn new information and make adjustments over time.

Different growers use different approaches to making the transition. Some prefer to start first by going organic on a small portion of the farm or ranch, whereas others prefer to make the transition on the entire farm at once. The choice really depends on the grower's preference, the particular vineyard's characteristics, and market conditions; there is no one strategy that works for everybody. Here are some examples of different approaches:

- **One parcel or block at a time.** Some growers start the transition on a limited acreage and choose a parcel size based on finances and labor availability. They prefer to certify one portion or block of the farm at a time, to begin by experimenting, and to minimize risk. If there are added economic costs during that transition period, those costs can be minimized on a small portion or block of land.

- **Gradual adoption of organic methods.** Other growers choose a gradual approach, adopting organic practices one by one over time on all or part of the farm. For example, they may start by using

continued on next page

continued from previous page

cover crops at first, then using compost instead of chemical fertilizers the next season, then stop using certain herbicides, and then stop using insecticides, etc. This approach might take a little longer, but some people prefer it as a way to spread the costs and changes over time.

- **Whole-farm "immersion."** Finally, still other growers choose to make a more rapid transition by using organic practices on an entire farm all at once rather than taking a step-by-step approach. Some people might view this approach as more risky, but it can have advantages because an all-at-once transition can enable a grower to make a significant change in the overall agroecosystem. The whole farming system may work more smoothly when it is not broken up into organic parcels and parcels that are still conventionally farmed. Furthermore, if the grower is already accustomed to using some organic practices, this kind of switch may make more sense than doing it piecemeal.

Growers who are making the organic transition will certainly face challenges. However, those challenges can also be seen as opportunities to improve the overall success of the farm enterprise. Some of the common challenges a transitioning grower might come up against include the following:

- **Information.** In some areas growers may have a hard time finding information about organic practices and regulations. Fortunately, however, information resources on organic practices are becoming more and more widely available, including extension agents, bulletins, journals, and websites with technical details. Many of these resources are mentioned in this book. Some areas offer courses and seminars about organic practices. Certification agencies can be a good source for basic information about organic rules and methods.

- **Research.** Research on organic practices is lacking nationwide and worldwide, compared to conventional agricultural research. Most agricultural researchers have been trained to focus on specific disciplines and a reductionist approach to solving problems. Nevertheless, a growing number of research scientists are addressing organic issues and the systems approach in recent years. USDA is slowly increasing its support for organic research via competitive grants offered through the Sustainable Agriculture Research and Education (SARE) program and other grants programs within the Cooperative State Research, Education, and Extension Service (CSREES).

- **Time management.** Managing time and resources may be a challenge on any farm, whether organic or conventional. But keeping track of organic inputs and practices is particularly important and can be time consuming, especially during the initial years of transition. Gaining new skills and knowledge also may take added time. Even so, good record keeping and time management skills are valuable in any farming operation, and going organic gives growers an opportunity to improve in these areas.

- **Marketing.** Organic winegrape growers sometimes encounter difficulties when marketing their product to buyers. Many wineries do not offer any premium or recognize any added value for organic grapes. Therefore, it is important for transitioning growers to make significant efforts at finding market opportunities. Fortunately, growing numbers of wineries and winemakers are becoming interested in organic winegrapes, partly because of an increasing recognition that grapes that are grown organically can make better-quality wines.

farm for certification. This is not required and adds to the transition's cost, but a good consultant could help the grower avoid mistakes that might otherwise slow the farm's progress toward certification. Upon obtaining certification, the grower must implement all production practices addressed in the regulation and must be proactive in building up soil organic matter with compost, cover crops, or other amendments, using organically acceptable methods to control weeds, insects, diseases, and any other pests, and implementing biodiversity. The grower must document in detail the name and amount of every material applied to the field and must keep the OSP current. A producer does not need to be certified or registered during the transition period.

Keeping Records from the Start

It is very important that growers keep careful records of the exact date they began the organic transition, the last prohibited materials applied to the field and when they were applied, and all of the specific inputs and practices used during the 3-year transition period. When you apply for certification, the certifier will need detailed production practice records in all of those areas. We recommend that you keep receipts and labels of organic inputs and that you get verification from neighbors, PCAs, county agricultural commissioner's pesticide use reports, or other local officials to corroborate that you ceased using prohibited pesticides (and when you did it) and when you started the organic transition. Also, it is helpful to keep records of production activities or tasks by recording work orders or maintaining other labor records.

Contacting the Certifier

The certifier will send you all of the application materials you will need to get started. Certifiers expect growers to read the handbooks or literature they provide. The grower's OSP must be fully prepared and submitted with the appropriate forms. Many certifiers require that growers sign an affidavit that shows land use history and affirms the truthfulness of the application. It usually takes 2 to 5 months for the application to be reviewed. Certifiers charge certification fees, which generally include

- a one-time application fee (ranging from $100 to $300)

- inspection costs, which usually pay for an inspector's time to travel to and inspect the farm and prepare a report, usually once per year

- annual fees (some certifiers use a fee based on acreage; others charge an annual fee of some set percentage of the gross production value)

Make sure you know the specific fees that the certifier will charge before you apply. The application fee must be paid when you submit the application. The fees cover the costs of the certification procedures and agency overhead and pay for the time of the inspection staff. Through certification, growers gain transparent third-party verification of their compliance with NOP. This does open the farming operation to greater public scrutiny, but ultimately it enhances the value of the farm's products in the marketplace.

Review of Application, Inspection, and Notification of Certification

The certifier will review the grower's application and inspect the operation before certification is approved. First, a trained inspector calls the grower to set up the initial inspection. After the inspection, the inspector submits a report to the certifier for review. The certifier informs the grower of the certification status or lets the grower know about any further requirements the grower must meet to achieve or maintain certification.

During an inspection, the inspector will want to review or see records about

- land use history

- pesticide use reports

- cleaning or cleanout procedures if you use shared equipment

- your understanding of the regulations (e.g., indication that you have read the appropriate manuals)

- the use of approved materials only

- appropriate systems to ensure compliance

- product labels for products or inputs that you use

Annual Review and Inspection

Organic certification inspections are done on farms once a year as required by the NOP. The inspector generally informs the grower of each visit in advance and sets up an appointment for the inspection. Certifiers and state organic inspectors do, however, have the right to make unannounced visits or inspections as well. If an inspector encounters violations or problems, the grower will receive a notice of minor noncompliance or major noncompliance from the certifier. If the noncompliance is minor, the grower is reprimanded, told not to repeat the error, and may be sent a reminder about the particular issue. If the inspector considers the noncompliance to be major, the grower's certification status may be affected. The grower may be required to submit requested documents or update a section of the OSP, or in serious cases the grower can be decertified.

As discussed previously, certification inspectors are not allowed to provide advice or information to the grower while conducting an inspection. They do not serve as consultants or advisors. This is intended to ensure that the certifier's role is one of enforcement and compliance checking. However, some certification agencies do have departments or divisions that provide information or educational materials to all growers, separate from the certification enforcement. In addition, many organizations, services, advisors, websites, and other resources provide information about organic practices.

A summary of the certification process and the interaction between the grower and the certifier is outlined in Table 3.3.

Table 3.3. Description of roles and responsibilities of the certifier and the grower

Grower's responsibilities	Certifier's responsibilities
Transition period:	
Read and understand National Organic Program (NOP) regulations. Gather information. Do research and planning. Begin transition.	
Document practices and management for 3 years.	
Select an accredited certification agent.	Certifier only has a role when the grower seeks transitional certification.
Request application materials and instructions from certifier.	Provide materials and information to client (e.g., NOP regulations, application materials, etc.).
After 3 years:	
Register with CDFA prior to sale of organic products.	Certifier is not involved in registration; the grower registers through CDFA.
Submit application and organic system plan (OSP) to the certifier for certification.	Verify that the grower is registered with CDFA and review the application for certification in timely manner. Request additional information from grower as necessary.
Supply any new information or details requested by the certifier to demonstrate compliance. Grower must submit to on-site inspection.	Must conduct inspection and notify grower of results. Upon approval, the OSP becomes a legally binding contract.
Implement the OSP. Keep records of practices and procedures. Obtain prior approval for any alterations to the OSP. Keep accurate and detailed records, and keep them organized in preparation for annual inspection.	Annual review and inspection to ensure compliance with terms of certification.

Issues Regarding Compliant and Noncompliant Materials

All materials that can be used in crop and livestock production are classified into one of four categories under the NOP. Note that in the context of the NOP regulation, the words "nonsynthetic" and "natural" are used synonymously.

- **Category 1: Allowed nonsynthetic.** Most nonsynthetic (natural) materials are allowed, with the exception of those that are specifically prohibited (category 2, below). There is no list of allowed natural materials; the rule of thumb is that natural materials are allowed unless they are specifically prohibited.

- **Category 2: Prohibited nonsynthetic.** A few nonsynthetic (natural) materials, such as arsenic and strychnine, are not allowed under the NOP. See NOP Rule sections 205.602 and 205.604.

- **Category 3: Allowed synthetic.** These synthetics are allowed, but their use is restricted to specific purposes as defined in the NOP regulations and National List. Use of these materials for other purposes is prohibited. See NOP Rule sections 205.601 and 205.603.

- **Category 4: Prohibited synthetic.** Most synthetic materials are not allowed. The exceptions are those specifically approved for use in organic production (category 3, above).

The National List of allowed materials for organic production is overseen and maintained by the NOP (www.ams.usda.gov/AMSv1.0/nop). The National List indicates generic compounds and materials that are allowed but does not list specific brand-name products. Brand-name information is available through the Washington State Department of Agriculture (http://agr.wa.gov/FoodAnimal/Organic/

MaterialsLists.aspx) and the Organic Materials Review Institute (OMRI) (www.omri.org). OMRI is a private nonprofit organization that provides verification and listings of products that meet the national organic standards. They publish and regularly update two lists: a brand-name list and a generic list.

As a certified grower, you are expected to keep track of updates and changes in the approval status of the materials that you use. Before trying new materials, you should research the NOP, OMRI, and WSDA lists to make sure that the materials are allowed and also get permission from your certifier to amend your OSP and use the product. Do not rely solely on the verbal or written declarations of vendors. Consult a reliable source to confirm any vendor claims.

The certification agency has the final decision on the acceptability of inputs for each farm operation. In making a determination about the acceptability of an input, the certification agency must evaluate the input based both on its ingredients and on the context in which it will be applied. Prior to using any input on an organic farm, you must obtain written approval (e.g., a certificate of compliance) from your certifier. The certifier will contact you and or the material's manufacturer if he or she needs any additional information to determine whether use of the material is in compliance. A word of caution: Any input that you use in your farming operation without the prior written approval of the certifier could be viewed as a departure from your agreed upon OSP and could be grounds for adverse action on the part of the certifier, USDA, or state programs.

CERTIFICATION ISSUES FOR WINERIES

Winegrape growers and wineries that wish to use an "organic" label for their product must be certified as complying with NOP regulations. The regulations state that "organic products must not contain… sulfites, nitrates, or nitrites added during the production or handling process," with the explicit exception that "wine containing added sulfites may be labeled 'made with organic grapes'" (NOP, 205.301(f)(5)). As indicated previously, if you are both a grower and a processor, both aspects of your operation need to be certified and need to submit to a yearly inspection by an accredited certifier.

Wine may be labeled as "100% organic" if the grapes used are certified organic and if all processing inputs and processes are in compliance with the NOP. If any nonorganic ingredients are added, the product may still be labeled as organic as long as the nonorganic ingredients constitute less than 5 percent of the final processed product. In both cases, the "USDA organic" label may be used, though it is not required.

To use the "organic" label, a wine producer must use grapes that are certified organic and must follow NOP regulations in the processing of the grapes for wine.

There are two options for labeling wine:

- "Organic wine" (only certified organic grapes are used and no added sulfites are used in the wine-making)

- "Made with organically grown grapes" (only certified organic grapes are used, and added sulfites are kept below 100 ppm)

Your goal as an organic processor is to maintain the organic integrity of the product. There are a number of critical points that organic wine producers may want to address in the organic handling plan for their operation. These include

- **Sanitation.** When you sanitize equipment that comes in contact with organic product, the sanitizer must be in compliance with NOP. Before you use the sanitizer, make sure to get it approved by your certifier. This information also must be documented and recorded.

- **Cleaning.** During crushing, the hoppers and bins must be rinsed with water. Cleaning between batches of conventional grapes and organic grapes must be documented and recorded.

- **Tagging and tracking.** All certified organic grapes coming in to the processing facility must be specifically tagged, identified, and tracked to ensure they are kept separate from other grapes that are not organic.

- **Winemaking ingredients and processing aids.** If you are a grower, you should consult NOP Sections 205.605, 205.270(b), and 205.105 for lists of substances that are allowed as ingredients in or on processed products. Also, make sure you know the compliance status of any processing aid material: check with your certifier before you use it. As indicated previously, added sulfites are not allowed in the processing of organic wine. For wine labeled "made with organically grown grapes," you can use sulfur dioxide, but the sulfite level must stay below 100 ppm. Although the NOP may not be explicit about certain other additives, there are a number of conventional winemaking substances that would be restricted under the general guidelines that prohibit the use of synthetic (non-naturally occurring) materials. These include sodium or potassium metabisulfate, genetically modified yeast, diammonium phosphate, and polyvinylpolypyrrolidone (PVPP). There are also restrictions involving the use of carbon as a filtering aid, so make sure you consult with your certifier regarding that material, too.

- **Water.** Any water that you use must be potable. Chlorine materials are allowed, but residual chlorine levels in the water should not exceed the residual disinfectant limit as set in accordance to the Safe Drinking Water Act.

Consider doing a periodic assessment of the organic critical control points in your operation (e.g., transport trailers, washing devices, pressing equipment). The key questions to ask are: Where might you lose track of which product is organic and which is not? Where could organic product be cross-contaminated with non-organic product?

REFERENCES

Dimitri, C., and C. Greene. 2002. Recent growth patterns in the U.S. organic foods market. USDA Economic Research Service Publication AIB-777. ERS website, http://www.ers.usda.gov/Publications/aib777.

Klonsky, K., and K. Richter. 2005. A statistical picture of California's organic agriculture: 1998–2003. UC Agricultural Issues Center Web site, http://aic.ucdavis.edu/research/StatisticalReview98-93f8.pdf.

Klonsky, K., and L. Tourte. 1998. Statistical review of California's organic agriculture: 1992–1995. UC Agricultural Issues Center Statistical Brief No. 6 (May). AIC website, http://aic.ucdavis.edu/publications/oldanrpubs/organicstatistical.pdf.

———. 2002. Statistical review of California's organic agriculture: 1995–1998. Oakland: University of California Agriculture and Natural Resources Publication 3425. UC Agricultural Issues Center website, http://aic.ucdavis.edu/research/misc/Organic1995-98.pdf.

OTA (Organic Trade Association). 2007. OTA's 2007 manufacturer survey. Summary at the OTA website, http://www.ota.com.

INTERNET RESOURCES

ATTRA (National Sustainable Agriculture Information Service), http://attra.org/organic.html.

California Organic Program (Processors), www.dhs.ca.gov/fdb/HTML/food/organicreg.htm.

California Organic Program (Producers and processors), http://www.cdfa.ca.gov/is/i_&_c/organic.html.

OrganicAgInfo (Scientific Congress on Organic Agricultural Research), http://www.organicaginfo.org.

Organic Farming Compliance Online Handbook, http://www.sarep.ucdavis.edu/organic/complianceguide.

Organic Materials Review Institute, http://www.omri.org.

USDA National Organic Program, http://www.ams.usda.gov/AMSv1.0/nop.

Washington State Department of Agriculture Organic Food Program, http://agr.wa.gov/FoodAnimal/Organic.

PART 2

Soil Management in Organic Vineyards

PHOTO · TOM LIDEN

Soil Management in Organically Farmed Vineyards

GLENN T. McGOURTY AND JOHN REGANOLD

Together, site selection and soil management form the foundation of high-quality winegrape growing. The best vineyard sites naturally possess well-structured soils that have adequate (but not excessive) soil fertility and soil moisture, providing a good rooting environment. In these locations, vines grow in a well-balanced manner, having a large enough canopy to ripen a moderate-sized crop that makes excellent wine year after year. Often, these soils contain mixed alluvium and colluvium, are well aerated, and have a stable soil structure formed from parent materials that give the soil good inherent chemical and physical properties. Unfortunately, this kind of a vineyard site is not commonly found. By applying thoughtful soil management practices, though, winegrowers can create these conditions even in sites that are less than ideal. An organically farmed vineyard requires adequate fertility, good physical and chemical conditions for roots, sufficient water and air for the roots to thrive, and protection from erosion. Experience has shown that organic practices can provide for all of these needs. The overall approach is different from that of conventional farming, but many conventional winegrowers use some of the same tools and methods as organic winegrowers.

Here are the objectives of soil management in organic winegrowing:

- Increase levels of soil organic matter (SOM).

- Improve soil physical conditions.

- Provide adequate nutrition using naturally occurring substances.

- Increase the number and diversity of soil organisms.

- Protect the soil from erosion and exposure to sunlight.

- Address imbalances in the soil that interfere with root growth and nutrient uptake.

What a Grapevine Needs

Plants use their roots to provide for several essential physical and chemical needs. The roots must be able to penetrate the soil in order to provide support for aerial portions of the plant and to find nutrients and water for growth. The soil itself must have a reserve of water sufficient to meet the plant's transpiration needs during the growing season. (California vineyards, because of the state's semi-arid climate, may require irrigation to augment the soil's water reserves.) Soils must be porous enough to allow air to move readily into the root zone, since roots will not grow without oxygen. Finally, the soil must contain adequate plant nutrients for the plants' roots to grow. The rooting environment must also be free of toxic elements (such as excess sodium, chlorine, nickel, and aluminum) and have a pH level that allows soil nutrients to be available to plants.

Soil Physical and Chemical Conditions

Because roots are beneath the surface of the soil, it is hard to see how they grow and what happens to them over the course of the growing season. Our understanding of vine root growth is surprisingly limited. A healthy root system is essential for vine performance, since water and nutrient uptake are key processes for quality fruit production, and this is completely dependent on the root system. Foliar feeding is an option, but it is not the most efficient way for a plant to absorb nutrients.

One priority in organic growing is to get your plants to develop an extensive root system that utilizes a maximum area of the soil surface for water and nutrient uptake. Often, soluble nutrient levels in the rooting zone are lower for an organic farming system than for a conventional farming system, since concentrated chemical fertilizers are not allowed under organic certification (some concentrated organic fertilizers are available, but they are quite expensive and their use is limited). To compensate for this, organic growers have the goal of growing a larger root system that will contact a larger soil volume. The more extensive root system contacts more of the soil so it can absorb more nutrients, even though they are present in lower concentrations than in conventional farming systems. Organic growers depend on both the mineral and organic fractions of the soil to provide nutrition to the vines (they sometimes use foliar fertilizers as well, but usually only to supply micronutrients).

Another goal in an organic farming system is to build up higher levels of soil organic matter than are normally found in conventional farming systems. The minerals in the organic fraction of the soil are released slowly, unlike soluble chemical fertilizers, which usually become available immediately upon application. In an organic farming system, the release of nutrients for use by plants often depends on other biological processes in the soil. Since most organic fertilizers are less soluble than most conventional fertilizers, it is less likely that soluble nutrients will leach from the soil profile and end up polluting surface water or groundwater, a common hazard with many of the highly soluble chemical fertilizers used in conventional agriculture. It is important to note that under some conditions excessive nutrient leaching is possible in organic farming as well, though the practices that organic winegrowers use are not likely to cause this sort of problem.

High root density and long root hairs are necessary prerequisites for a vine's efficient uptake of nutrients from the soil, and increased root density improves nutrient uptake. The density of a vine's root system is determined in part by rootstock genetics: species differ in their root morphology. The other limitation determining root density is the soil's physical characteristics. Roots will not grow well or at all in a soil profile that has low porosity or a low oxygen level or one that is hard, dry, infertile, or toxic due to mineral excesses (sodium, aluminum, manganese, nickel, and hydrogen) sometimes found under particularly high or low soil pH conditions.

Plant nutrients are transported in the soil by diffusion through the soil solution, by mass flow of the soil solution through the pore system, and by root interception of soil nutrients. Mass flow of nutrients toward the roots is faster than diffusion

and can have an effect over a larger distance. Nutrients moved by mass flow include sulfur (S), calcium (Ca), sodium (Na), magnesium (Mg), the nitrate form of N, and boron (B). Due to the low concentrations of potassium (K), phosphorus (P), nitrogen (N), copper (Cu), iron (Fe), manganese (Mn), and zinc (Zn) in the soil solution, though, mass flow cannot address plants' needs for these elements. Instead, they have to be augmented by diffusion. This means that the roots have to be in close proximity to the sources of these nutrients. Well-structured soil with low bulk density and high porosity allows roots to explore the soil profile more fully, and in turn create a larger root system and intercept more nutrients than a limited root system growing in a compacted soil (Cass 2005). This is one reason why compost and cover cropping are important tools for solving micronutrient deficiencies in organic winegrowing. Compost also contains micronutrients that are available to the grapevines. Organic matter added to the soil can chelate micronutrients, making them more available than they would be in soils that are low in organic matter. All of these factors help improve micronutrient uptake.

When a soil becomes compacted, less porous, and hard, the availability, movement, and uptake of potassium and phosphorus are negatively affected. These nutrients are only available to roots in soils that are well structured and adequately moistened. As soils become compacted, all diffusion-based nutrients (potassium, phosphorus, nitrogen, copper, iron, manganese, and zinc) become less available.

Mycorrhizae

"Mycorrhizae" is a term for the associated mass of the hyphae of soil-dwelling fungi and the roots of vascular plants on which the fungi live. The two organisms have a symbiotic relationship: the fungi receive nutrition from the vascular plant, and they also help the host plant to take in nutrients and water from the soil. Numerous studies have shown that healthy grapevines are dependent on mycorrhizae and that the vines grow poorly in their absence. In vineyards, the fine roots of grapevines can form mycorrhyzal associations with fungi from the order *Glomales*. Researchers have clearly shown that these arbuscular mycorrhyzal fungi (AMF) assist in the vines' uptake of phosphorus, and they may also help with nitrogen, potassium, and zinc uptake. AMF may also improve the vines' resistance to drought, and they appear to colonize vine roots readily under dry soil conditions.

Balancing Soils

An evaluation of a soil's cation-exchange capacity (CEC) and the ratio of elements in the base saturation of the CEC can help you predict potential problems in soil structure and nutrient availability. The cation-exchange capacity of a soil is the total quantity of cations (positively charged ions, pronounced "cat ion") that a soil can absorb by cation attraction, usually expressed in cmols per kg of soil. The soil's cation-exchange sites occur on the surface of clay particles and organic matter. The more clay or organic matter is present in a soil, the higher its CEC will be (some clay minerals—such as montmorillonite, vermiculite, and smectite—also tend to have a higher CEC per unit volume). Higher CEC often indicates the potential for a fertile soil with good water-holding capacity, assuming that the soil has a good structure. Base saturation is the percentage of the total CEC occupied by basic (as opposed to acidic) cations, including Ca^{++}, Mg^{++}, K^+, NH_4^+, and Na^+. In acid soils, H^+ and Al^{3+} ions compete with the basic cations, diminishing the exchange and uptake of cations by plants. For a soil of any given organic and mineral composition, the pH and fertility level generally increase with an increase in the degree of base saturation (unless the base cations are Mg^{++} or Na^+).

In soils in which the base saturation is dominated by a large percentage of magnesium and sodium, the nutrient potassium (and sometimes phosphorus) is often deficient, particularly toward the end of crop ripening. Soils high in sodium are rare in the premium wine-producing regions, but soils high in magnesium are quite common in coastal vineyard regions, especially where the soils were formed from ultramafic rocks such as serpentinite and mafic rocks such as basalt (Anamosa 2005). Structural problems are likely to occur in these magnesium-rich soils: because of poor soil physical conditions, limited root growth, restricted amounts of oxygen, and poor potassium diffusion, vines will grow poorly. Large initial applications of calcium followed by smaller maintenance applications along with compost and cover cropping can greatly improve soil structure and vine growth in these sites. Periodic soil testing along with recommendations from a knowledgeable consultant can help you develop a program to address these potential problems. Normally, growers use gypsum for this purpose due to its solubility and its capacity to help leach Mg and Na from the soil profile. The potential to improve the soil with these treatments increases when additional organic matter and cover crops are integrated into the operation.

It is common practice among conventional and organic growers to apply either gypsum (for soils high in sodium and magnesium and having a neutral to high pH) or lime (for soils high in hydrogen and having a low pH) as a soil amendment on a regular basis to address imbalances in the soil's cation-exchange complex. It is a very good idea for a grower to review the need for calcium-containing amendments with a qualified soil scientist who can help design a program to help maintain soil structure and fertility. This is a very important part of an organic soil management plan.

Many crop consultants believe that bringing calcium levels up to a 65 or 70 percent base saturation will greatly improve soil physical conditions, create a favorable pH, and improve nutrient uptake of many minerals, especially potassium. Some also believe that Ca:Mg ratios of 5:1 are desirable for the same reasons. There is no question that soils that are high in magnesium and sodium will benefit from calcium applications, usually applied as gypsum. Similarly, acid soils in which the CEC is dominated by H^+ and Al^{3+} ions are improved when calcium is added (usually as lime), increasing the calcium proportion in the soil's base component. If a soil has a high clay content, this approach may require the application of large amounts of calcium soil amendments. In contrast, soils that are low in potassium, low in magnesium, and low in calcium (such as sandy, acidic soils) might actually become potassium deficient after large applications of calcium. Most growers work with a soil professional to review the soils report before they start on a calcium amendment program for their vineyards.

Starting an Organic Vineyard in a Previously Unplanted Area

When you are planning an organic vineyard on a site that has not been farmed before, it is advisable to do a thorough evaluation of the site, including sampling of the soil and water.

In developing a new vineyard that is to be farmed organically, growers often begin by using deep tillage to loosen the soil in the area where the vines will be planted. The purpose of deep tillage is to loosen the soil so that there are no limitations to the growth and extension of roots into the soil profile. Usually it is sufficient to loosen the soil to a depth of 3 feet. If done properly, deep tillage will reduce the soil's bulk density and increase its porosity. Excessive tillage should be avoided, though; soil aggregate destruction that results in a loss of soil structure can result when soils are overworked. Compaction will result when the soil structure is damaged by excessive tillage. If soil aggregates have been reduced to dust, tillage is clearly excessive. After such treatment, the soil is less porous and its bulk density is increased, creating a more difficult rooting environment for the grapevines.

Deep plowing and inversion of the entire soil horizon is not helpful in most situations and can be quite harmful to the soil's structure and general health. Rather, deep tillage should be done with shanks that have been modified with a sweep at the bottom to fracture and displace the soil in an upward movement, breaking blocky or prismatic soil structures into smaller aggregates that will allow better aeration, water penetration, and biological activity to improve soil structure and porosity, which will in turn soften the soil and allow root penetration. In California, the time for deep tillage is late summer and early fall when the soil is slightly moist but not dry. Under these conditions, soils are more likely to fracture into smaller aggregates. If worked too wet, though, the soil will simply smear when penetrated and will actually end up more compacted. If worked too dry, excessive tillage will cause the soil to turn to powder, and small aggregates will be destroyed (Cass 2005).

Starting an Organic Soil Management Plan in an Existing Vineyard

When you are converting a vineyard that is already planted and has been growing under nonorganic management, shallow ripping where tractor wheels run can be useful if compaction in this area is evident. If the vineyard has been cultivated, you may also want to evaluate whether a plow pan exists in the row middles at the depth of penetration of the tillage equipment that has been used there.

You can use short shanks 10 to 12 inches long with small sweeps to loosen soils where compaction is a problem. This usually is done in the fall, when compost, lime, or gypsum is applied and winter cover crops are seeded. A drawback to this step is the potential for damage to shallow grapevine roots. In most cases this will not be a problem, but make sure to shallow-rip a small area first and evaluate it to ensure that you are not damaging vine root systems before you try this treatment for the entire vineyard.

By following these specialized tillage operations, applying compost and other soil amendments, and growing cover crops, you will help stabilize the newly created soil porosity and help increase the formation of new pores and stable aggregates in the soil. A reduced soil bulk density and increased opportunities for root exploration of the soil profile are the ultimate goals of these practices. Another benefit is that water infiltration rates and storage capacity will increase, and that lessens both the potential for soil erosion and the need for irrigation.

Increasing soil organic matter helps in the long term to keep the soil from compacting easily, and soils with stable aggregates are more resilient following wheel traffic and other activities that can cause compaction. It also makes sense to restrict or prohibit the use of heavy tractors and other vehicles in the vineyard during periods of high soil moisture, when soil compaction is most likely to occur. Vehicles with low ground pressure such as tracklayer tractors or lightweight ATVs have obvious advantages in reducing soil compaction during normal vineyard operations.

BUILDING SOIL ORGANIC MATTER (SOM)

In this section, we will consider the theoretical aspects of how soil organic matter (SOM) is formed and managed. In organic farming, the building up of SOM is an objective because increases in SOM tend to improve soil physical properties and, often, the availability of nutrients for plants. By adding organic matter, you help both directly and indirectly to improve the structure of the soil. Soils with granulated aggregate structure are far less erodible and serve as a better rooting environment than soils that are structureless and massive (Brady and Weil 2002). Frequent tillage destroys aggregates and can cause massive, structureless soil, resulting in a poor rooting environment (Baver 1956).

Usually, organic matter is added either as compost or as cover crops. As mentioned earlier, organic matter stimulates soil biological activity, which is extremely important in helping to improve soil physical conditions, including

- reducing soil bulk density by increasing soil porosity

- reducing soil resistance to allow roots to extend easily

- improving infiltration of water and air

By definition, organic matter is living or dead plant and animal tissue in the soil. Compost applications and cover crops should help to increase SOM, as plant tissue is the primary source of SOM. Typically, there are three interrelated organic fractions in the soil (Figure 4.1):

- biomass (1 to 8%), made up of living organisms (averages 3% of SOM)

- plant and animal residues (10 to 25%) in various stages of decomposition

- humus (60 to 70%), the semi-stable fraction of SOM remaining after the major portion of added plant and animal residues has decomposed

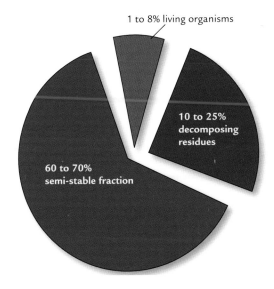

1 to 8% living organisms

10 to 25% decomposing residues

60 to 70% semi-stable fraction

Figure 4.1. Organic matter fractions making up a typical soil.

Soil organic matter serves many purposes. First, polysacchaccarides and other beneficial metabolites are created by soil organisms as they digest organic matter. Microbial products act to bind soil particles together to improve soil structure (smaller aggregates and soil particles are bound together, forming larger aggregates). Larger voids (macropores) form between the larger aggregates, increasing the soil's porosity and allowing the rapid movement of air and water. The macropores are often formed by and associated with worms and other soil-dwelling invertebrates (some soil physicists call these voids biopores). Worms feed on decaying plant tissue and help to create stable soil aggregates, known as worm castings, that have passed through the worms' digestive system. Plant roots growing through the macropores can also expand the pores as they grow, helping to increase the overall porosity of the soil. As the number of macropores increases in a soil, the soil's bulk density and its strength are reduced, making for a more favorable rooting environment (Baver 1956).

An increase in SOM may help improve a soil's water-holding capacity. Also, improvements in soil porosity and reductions in its bulk density help plant roots to fill the soil profile. These improvements help reduce grapevine stress and may also reduce the need for irrigation (Ingels et al. 1998).

Humus is composed of humic and fulvic acids, large molecules that can significantly increase a soil's cation-exchange capacity. When these humic substances bind together, they create larger organo-mineral assemblages that contribute to the soil's structure and increase its water-holding capacity. These effects are particularly noticeable in sandy soils. Heavy clays can also benefit dramatically, becoming more friable. Humus is the end product of microbial decomposition. It appears as a darkening layer that coats soil particles. In an agricultural system, this layer can help protect groundwater from contamination by pesticides, an overabundance of some nutrients, and toxic heavy metals. Higher levels of SOM also help buffer the soil against a rapid change in pH (Gardiner and Miller 2004).

Grower experience shows that it takes 3 to 5 years of a soil-building program (compost applications and cover cropping) before the SOM levels begin to make noticeable differences in a vineyard's growth and the quality of the soil. It makes sense for it to be a long process, given that so much of the organic matter you add to the soil is respired and returned to the atmosphere as CO_2. The accumulation of stable humus simply takes time, as do the effects of this material on the soil and on vine growth. Soil tilth, adequate fertility, and soil biological activity are indicators of change when you make an effort to accumulate more organic matter in the vineyard soil. Still, there can be very noticeable short-term benefits:

- protection from soil erosion

- increases in soil nitrogen and vine growth in low fertility sites

- competition with vines in vigorous sites (helping to reduce vine vegetative growth)

- suppression of weeds, depending on cover crops planted and management procedures utilized

Factors Affecting the Accumulation of SOM

As cover crops decompose (especially the roots, since they are already well distributed in the soil), SOM is formed. The rate of SOM accumulation will depend on several factors:

Climate. Soil organic matter accumulates most rapidly in cool, humid regions. In a grassland prairie in the northern Midwest, the rate of SOM accumulation would be near its peak. In contrast, SOM accumulation is at its lowest in warm, dry places. Biological activity is significantly reduced under low moisture conditions, and organic matter is prone to photodecomposition if exposed to intense sunlight and prone to oxidation if it is exposed to the atmosphere. Desert soils typically contain less than 0.5 percent SOM. Since California vineyards are in warm, sunny, dry Mediterranean climates, building SOM there can be a challenge. Just through the use of organic farming practices, though, a California vineyard's soil can accumulate as much as 2 percent SOM, or even more. It is easier to accumulate SOM in areas where the climate has long periods of cool, humid weather (Brady and Weil 2002; Ingels et al. 1998).

Soil texture. Accumulation of SOM is generally highest in clays and silt loams and lowest in sands (Plaster 2003). In general, finer-textured soils are more fertile, have a greater water-holding capacity, and have more total surface area on soil particles available to bond with humic substances. It is possible to build SOM content up to 3 percent

or more in vineyards with fine-textured soils. It is much more difficult to build SOM beyond 1.5 percent in coarse-textured soils, but even this level can still have a positive impact on overall soil quality (Brady and Weil 2002).

Tillage. The rate of accumulation of SOM varies depending on the amount and intensity of tillage in a given field. Although cover crops certainly stimulate biological activity and respiration, tillage causes a considerable amount of organic matter in the soil to decompose, sometimes leaving little behind. Exposure of SOM to sunlight and the atmosphere also hastens its decomposition. Soil erosion is also higher when tillage leaves bare soil exposed to rainfall, and SOM washes away with the erosion. Conservation tillage practices minimize the displacement of soil and leave materials on the surface of the soil profile to maximize SOM accumulation (Brady and Weil 2002; Chapman 1985; Ingels et al. 1998). There are many benefits to employing a no-till vineyard floor management system, including prevention of erosion, creating habitat for beneficial insects, and improving water infiltration in the vineyard (see chapter 6, Cover Cropping Systems for Organically Farmed Vineyards).

Cover crops and organic matter additions. The amount and quality of the organic matter you add to the soil will influence the rate at which SOM develops. This is discussed in some detail later in this chapter.

Compost, Cover Crops, and Soil Biological Activity

Soil microbes perform numerous soil functions, such as nitrogen and carbon cycling, decomposition of plant residues, and formation of SOM and soil aggregates. Organic winegrowers apply many nutrients either in the form of naturally occurring compounds or in organic materials that require further decomposition. The rate of release of these nutrients is dependent on soil microbial activity. Compost applications, cover crops, and associated management practices can impact both the composition and the activity of microorganisms in the soil. A number of scientists have noted improved soil biological activity and soil quality where these practices have been applied (Ingels et al.1998; Morlat and Jacquet 2003; Reganold 1988).

High levels of soil biological activity are frequently listed as indicators of soil vitality (Table 4.1). In general, the addition of organic matter stimulates biological activity in the soil. Organic matter serves as an energy source for the ecological community in the root zone, known as the rhizosphere. Soil with vegetation supports higher microbial populations than fallow soil. While they live, plant roots exude compounds such as amino acids, simple sugars, and organic acids and shed cells that contain polysaccharides. These compounds provide a continuous energy supply to microorganisms living in the root zone. Soil microbe activity closely follows the patterns of active root growth of plants. If the cover crop is mowed but not tilled, the cut portions of the plant provide habitat, shade, and food at the soil surface. If the cover crop is plowed into the soil and soil moisture is favorable, all of the plant parts become available as a food source. There will be a flush of soil microbe activity until the cover crop residues are utilized, and then microbe populations will decline (Ingels et al. 1998).

How Cover Crops and Organic Matter Decompose in Soils

One of the major effects of a cover crop on the soil is the increased organic matter contributed by cover crop residues (Table 4.2). A vineyard yields relatively little biomass to return to the soil during the course of a growing season, averaging between 1,700 to 2,800 kilograms per hectare of prunings (fresh weight) and 900 to 1,700 kilograms per hectare of leaves (dry weight). Cover crops can increase these crop residue additions to a vineyard by 2,000 to 4,000 kilograms per hectare of dry matter. Addition of this material stimulates all biological activities in the soil, especially if you use legume cover crops. Plant tissue is the primary source of organic matter in soils. Fresh herbaceous plant material is 60 to 90 percent water by weight. Between 90 and 95 percent of plant dry matter consists of carbon, oxygen, and hydrogen. The remainder includes potassium, sulfur, nitrogen, phosphorus, and other nutrients (Ingels et al. 1998).

When organic residues are added to soil and there is an adequate amount of nitrogen, the entire biota (the population of living organisms) of the soil increases. The microbes incorporate much of the nitrogen from the residues into their bodies. Nitrogen in this form is fairly stable, making it unavailable to plants. Then nematodes, earthworms, and protozoa consume the microbes, using the carbon compounds for energy, and excrete much of the nitrogen (the portion that they cannot utilize) in the form of ammonium ions, which can be assimilated by some plants. Nitrifying bacteria

Table 4.1. Beneficial activities of microorganisms in the soil

Function	Type of organism
Decomposition of plant residues	decomposers (bacteria, fungi, microfauna, and macrofauna)
Nitrogen availability to plants:	
Conversion of organic to mineral forms	mineralizers
Conversion of ammonium to nitrate	nitrifying bacteria
Denitrification of nitrate to nitrogen gas	denitrifying bacteria
Increase availability of phosphorus, iron, sulfur, and other elements	many organisms, mycorrhizal fungi
Formation of soil humus	all soil organisms
Soil aggregate formation and stability	polysaccharide-producing fungi and bacteria
Production of plant growth-promoting substances	primarily bacteria
Suppression of plant pathogens	general microbial community, producers of toxins, producers of iron chelators
Breakdown of pesticides and other toxic chemicals	biodegrading bacteria and fungi

After K. Scow (Ingels et al. 1998).

quickly convert this to nitrates, which can be utilized by the rest of the plants in the field. The process is temperature and moisture dependent but frequently also coincides with conditions favorable for plant root growth (Brady and Weil 2002; Ingels et al. 1998). Warm soil temperatures (67° to 80°F) and adequate soil moisture (near field capacity) allow the most rapid decomposition of fresh organic matter. The carbon-nitrogen (C:N) ratio of the cover crop is also important in the digestion and breakdown of cover crops by soil microbes. When C:N ratios are below 25:1, a condition common to plant material before flowering, the cover crops break down readily when incorporated into the soil. If the cover crop is allowed to mature and dry, C:N ratios may exceed 50:1 as the cellulose content increases, and subsequent breakdown of this material will be much slower when it is incorporated into the soil. Nitrogen may actually become unavailable to the perennial crop while this biological digestion is occurring. Consequently, most recommendations state that for maximum benefits and under a tilled vineyard floor management system, cover crops should be incorporated when in full bloom.

The organic compounds found in plants can be sorted into several broad classes. The following compounds are listed in order according to their ease of decomposition, with the quickest to decompose listed first. When cover crops are incorporated into soils, the following types of reactions are likely to occur with adequate moisture and moderate temperatures:

1. Relatively simple carbohydrates (sugars, starches) are enzymatically oxidized, producing carbon dioxide, water, energy, and being consumed by decomposition organisms.

2. Macro- and micronutrients are released and immobilized through a variety of pathways, most of them unique to each element.

3. Stable organic compounds form, either through modification of substances from the original decomposing tissues or through synthesis by decomposing organisms. These materials will become humus, organic glues and waxes, and other compounds that help to improve soil quality.

4. Proteins also decompose readily in the soil, yielding carbon dioxide, water, amino acids, and ultimately nitrate, ammonium, and sulfate ions that can be utilized by plants. This process, known as mineralization, releases elements from organic compounds into forms that higher plants and other microorganisms can use. Because legumes have high protein content, they can serve as a useful source of nitrogen.

The rate of organic matter decomposition is affected by the size of the decomposing particles and the placement of the material (Table 4.3). Coarser materials placed on the soil surface decompose slowly and require further decomposition and assimilation by soil-dwelling invertebrates and microorganisms. Smaller particles decompose much more quickly, particularly if they are incorporated into the soil (Brady and Weil 2002).

Table 4.2. Biomass cycling of conventional and organic/biodynamic farmed vineyards

Material	Amount under conventional farming	Amount under organic/ biodynamic farming
	------------- kg/ha -------------	
Compost	0	2,300 – 6,900
Leaves	1,400	1,400
Prunings	1,000	1,000
Weeds, cover crops	1,200	1,200 – 12,000
TOTAL (kg/ha)	3,600	5,900 – 21,300
TOTAL (ton/acre)	1.6	2.5 – 9.4

Cover Crop Mixes, C:N Ratios, and their Effect on Decomposition

Legumes are often planted together with a grass cover crop to biologically fix nitrogen while the cover crop is growing. When the crop is plowed into the soil, decomposing legumes ensure that there will be adequate soil nitrogen to assist in the decomposition of carbon-containing compounds produced by the grasses and forbs, as well as by the legumes. Legumes usually have a C:N ratio of less than 20:1, whereas grasses and many other forbs have C:N ratios of greater than 20:1. These ratios vary based on species and the plant's stage of maturity when incorporated; in general, the ratios are lower before flowering and increase rapidly as seed matures. The combination of grasses and legumes in the cover crop seed mix ensures that the cover crop's decomposition will be fairly rapid once it is incorporated into the soil. A typical cover crop mix for tilled vineyard floor management in California's North Coast area is bell beans, common vetch, winter peas, and barley. Several clovers, including subterranean, rose, and crimson clovers, are commonly planted in non-tilled vineyards (Ingels et al. 1998), along with annual grasses such as Blando Brome and Zorro fescue.

Carbon is a basic fuel for soil microorganisms, and they utilize this element in chemical compounds both as an energy source and as a building material for cell walls. Carbon-containing compounds alone are not sufficient to meet the microorganisms' growth needs, though. They need nitrogen and other elements to build proteins, enzymes, DNA, and other compounds essential for life.

In general, microbes living in the soil use about eight parts of carbon for about every one part of nitrogen, a ratio reflected in the average C:N ratio of 5:1 to 8:1 for soilborne microbes. Only about one-third of the carbon consumed by microbes is assimilated into their cells; the remaining two-thirds is used for respiration and lost as CO_2. This means that for every eight parts of carbon that are metabolized, 16 parts are given off as CO_2. Combine this with the 8:1 C:N ratio, and you see that soil microbes use 1 gram of nitrogen for every 24 grams of carbon. This ratio can vary, of course, and is dependent on the C:N ratio of the organic matter being digested.

Normally, if the organic material added to the soil has a C:N ratio that exceeds 24:1, the microbes will have to find nitrogen from other sources in the soil, and the rate of decomposition will be considerably slower. As the microbes utilize more and more free nitrogen in the soil, higher plants are subject to nitrogen deficiency. This is a common result when mostly mature grasses, which are high in cellulose, are plowed into the soil. Growers can use this phenomenon to their advantage as a technique to devigorate a vineyard that has too high a nitrogen level in the soil.

Cover crop selection. Grasses and sod-forming cover crops are the most effective for building SOM, since these residues are high in carbon and decompose slowly. The roots of these species die and are replaced seasonally during plant growth, and they

Table 4.3. Plant compounds and their decomposition in soil

Material	Source	Decomposition rate	End product
Sugars, starches	plant cell contents	fast	CO_2, H_2O
Proteins	plant cell contents	fast, moderately fast	N, P, S, CO_2, other inorganic nutrients
Hemicelluloses	plant cell contents	moderate to slow	CO_2, humus
Cellulose	cell walls	slow to very slow	CO_2, humus
Fats, waxes, oils	leaf surfaces, seeds	slow to very slow	CO_2, humus
Lignins, polyphenols	line cell walls	slow to very, very slow	CO_2, humus

Adapted from Brady and Weil 2002; Gardiner and Miller 2004.

are slow to decompose. Unfortunately, many grasses compete with grapevines for the same nutrients and water. Legumes decompose fairly rapidly due to their high nitrogen content, and so leave lower levels of residual SOM than grasses. Legumes may also stimulate grapevine growth by increasing the soil nitrogen.

Compost. Properly made compost contains relatively stable, long-lived organic matter similar to the humus that forms in soil. When you use good compost to fertilize cover crops, it increases SOM, especially in vineyards with active soil fauna. Favorable C:N ratios in compost also help with the degradation of incorporated crop residues.

Finally, compost applications to the vineyard serve as a source of nitrogen. Compost may stimulate the growth of cover crops if the compost contains significant amounts of nitrogen, phosphorus, and potassium. If incorporated into the soil, compost can contribute to the soil's nitrogen pool and help with the decomposition of carbon-containing crop residues. Compost C:N ratios range between 10:1 and 20:1. Carbon in well-made compost is fairly stable and contains humus that helps to build SOM (Brinton and York 2003a, 2003b). Compost is notoriously nonuniform: there are no commercial codes or regulations with regard to C:N ratios, nutrient content, or other standards that could help the end user to know what he or she is purchasing. Some composts made of recycled grape pomace and manure are very high in macronutrient content; other composts made from recycled green waste are low. Before you purchase compost, ask for a current lab analysis of the material.

Conclusion

Increased SOM is an important goal for an organic vineyard manager. Higher SOM levels can greatly improve soil quality and plant growth. Soil quality indicators such as biological activity, soil porosity, fertility, structure, water infiltration, and available water-holding capacity often increase as cover crops and compost applications help to build SOM levels. All of this in turn creates a better rooting environment, allowing roots to penetrate a larger area of the soil profile in search of nutrients and water. Many problem soils respond well to the practices used

in organic winegrowing, especially soils that start out with poor structure and chemical imbalances in their base saturation. As SOM increases, soil quality often improves fairly quickly, resulting in better vine growth and fruit quality.

DEVELOPING A SOIL FERTILITY PROGRAM FOR YOUR VINEYARD

In organic winegrowing, the grower needs to ensure that the nutrition needs of the vines are met so the vineyard will have an economically viable yield of fruit that has adequate nutrient content to avoid fermentation problems in the winery. When a new vineyard is being planned or an existing vineyard is undergoing transition to organic culture, preplant soil analysis is essential to determining whether there are any problems that need to be addressed. These might include unfavorable pH, poor Ca:Mg ratios, excess sodium, nutrient element deficiencies, or other conditions that would require either attention before planting or long-term management.

The approach known as soil building used by organic winegrowers to provide fertility differs from the conventional vineyard approach of adding soluble nutrients in that the results of the organic approach take more time and are not always as predictable. Soil fertility and vine nutrition under organic culture tend to reach plant nutrient level targets that are "adequate" rather than the "precise" levels targeted by conventional approaches. This is a weakness of the organic approach. Practical experience, however, indicates that most growers can develop a soil fertility program using organic methods that meets their vines' needs year after year. For an existing vineyard, organic growers use petiole analysis, visual observations, and fruit and pruning weight measurements to evaluate the effectiveness of their fertility practices. These are excellent methods for determining whether the vines have adequate nutrition and vigor. Problem areas where the vines are not growing well should be subjected to separate petiole and soil tests. They may require adjustments in the amount of fertilizer, soil amendments, and water to ensure a growth rate that matches that of the rest of the vineyard.

Essential Elements for Winegrapes

Winegrapes are well adapted to a wide range of soil types and fertility levels. They can perform surprisingly well in shallow and infertile soils. Many winegrowers deliberately seek out these sites, knowing that the concentration of flavor and pigments can be more intense if the vines have just their basic needs met for nutrition and water, without encouraging excessive vegetative growth. In general, grapevines have fewer nutrient element deficiency problems and require a lower level of fertility than most horticultural crops. Most winegrowers focus on providing adequate potassium and nitrogen for their vines. Other deficiencies are less common but if present can drastically affect crop yields and quality. Some nutrient excesses can also be harmful.

Sixteen elements are known to be necessary for normal plant growth: carbon (C), hydrogen (H), oxygen (O), nitrogen (N), phosphorus (P), potassium (K), calcium (Ca), magnesium (Mg), sulfur (S), zinc (Zn), boron (B), iron (Fe), manganese (Mn), copper (Cu), molybdenum (Mb), and chlorine (Cl). Carbon, hydrogen, and oxygen are taken in by plants from the air and water. The remaining 13 nutrients are absorbed from the soil by the plants' roots. These remaining nutrients are broken up into two groups: macronutrients (nutrients used in large quantities, expressed in pounds per acre) and micronutrients (nutrients used in small quantities, ounces per acre). Macronutrients include nitrogen, potassium, phosphorus, magnesium and sulfur. Micronutrients include zinc, boron, iron, manganese, copper, molybdenum, calcium, and chlorine.

Compost is an obvious source of both macronutrients and micronutrients. It is a cost-effective source for most nutrients, especially if the blend contains grape pomace and animal manures, since potassium levels are normally higher in such composts than in composts made entirely from green waste. You should recognize that nutrient element content of composts can be highly variable, making a lab analysis of the material essential to determining exactly what you are applying to your vineyard. Practical experience in the California's North Coast area has shown that annual compost application rates of 1 ton per acre often give adequate nutrition

to the vines when the grower also grows nitrogen-fixing cover crops. Compost use also has numerous side benefits to the environment and soil health, such as recycling a waste product that otherwise would end up in landfills, improving soil structure, increasing the amount and diversity of soil microflora and fauna, and improving the soil's water and air infiltration rates and its water-holding capacity.

Micronutrients are typically applied to grapevines as pre-bloom foliar sprays. Many products are available that have been organically approved, including seaweed extracts, amino acid chelated extracts, mined minerals, and other proprietary formulations of micronutrients. These products tend to be expensive, but application rates are generally very small. Compost provides a sufficient supply of many of them in the long term, but foliar sprays are often helpful for immediate relief of micronutrient deficiency problems.

The following tools and materials are available to help growers maintain soil fertility in the organic vineyard:

- *Cover crops* to fix nitrogen and cycle other macro —and micronutrients. There are many options for planting cover crops, but usually organic winegrowers plant legumes in at least every other row, hoping to generate 30 pounds of nitrogen per acre per year.

- *Compost* to provide macronutrients and micronutrients. Common application rates are 1 to 3 tons per acre per year, depending on soil quality and vine vigor. Compost is a very important source of N-P-K and micronutrients. Most growers will apply compost at least once every 3 three years. On fertile sites, a grower might find that cover cropping is adequate by itself, and compost application is unnecessary.

- *Naturally occurring concentrated organic fertilizers* made from biomaterials. Usually, these are used only in young vineyards that need extra nitrogen to stimulate vegetative growth. They may be applied either directly to the soil or through the irrigation system (fertigation).

- *Naturally occurring mined minerals to provide macronutrients and micronutrients.* Potassium sulfate and rock phosphorus are the most commonly used of these materials in organic vineyards. Applications can be fairly large if needed, and their effects may last for 3 to 5 years. (Rock phosphorus may fail to mineralize under high-pH conditions.)

- *Mined and naturally chelated micronutrients.* Depending on the material, these may be applied either as pre-bloom sprays or through the drip system. Zinc and boron are the micronutrients most likely to be required.

In general, these materials are initially more expensive than conventional soluble fertilizers, particularly when consider per unit of fertilizer. The advantage an organic soil fertility program has over a conventional one is that many of the organic fertilizers are slow release, and become available to the crop over time. After 5 years of a soil-building program, the need to continually add compost and grow nitrogen-fixing cover crops begins to diminish in vineyards that are building soil organic matter.

Stable soil organic matter contains a bank of nutrients that are slowly released over time by natural processes. These minerals do not solubilize readily, so they are less likely to leach and move off-site into surface waterways, a definite environmental plus. Many organic fertilizers are neither toxic nor caustic and can be handled and stored safely on the farm. Finally, many of the organic materials encourage the activity of soil organisms and biodiversity in the soil, and that can have beneficial secondary effects for root health, improved soil physical conditions, and an improved rooting environment (Table 4.4).

CALENDAR FOR MANAGING SOIL FERTILITY AND GRAPEVINE NUTRITION

October and November

1. Harvest is finished.

2. Apply compost, lime, or gypsum to vineyard, usually at the rate of 1 to 3 tons per acre,

Table 4.4. Nutrient budget for 1 acre of an organically farmed vineyard, North Coast (lb of nutrients per acre)

Inputs (source of minerals)	N	P	K	Ca	Mg
	------- lb/acre -------				
Composted pomace and manure	49	8	62	59	22
Rainfall deposition (EPA estimates)	11	0	0	2	
N from cover crops (half of rows, every other row)	10	0	0	0	
Soil tillage and mineralization from soil organic matter	10	1	5	5	
TOTAL INPUT lb/acre	80	9	67	66	
ESTIMATED AVAILABLE FOR CURRENT SEASON (Compost availability estimated at N=15%, K= 85%, Ca=85%)	38	2	57	56	
Outputs (based on 3-ton winegrape yield)					
Winegrapes	8	4.3	21	12	1.6
Trunk, stems, and leaves	15	1	1.8	11	3
Leaching and mineralization	15	0	5		
ESTIMATED TOTAL NUTRIENT REMOVAL	38	5.3	28	33	4.6

*Adapted from Brinton and York 2003.

depending on soil and crop needs. Usually, the material is applied to the entire vineyard floor. Winter rains will stimulate biological activity, including insects and earthworms, which will feed on the compost. Microbial activity also begins to consume and incorporate the compost into the soil.

3. Take soil samples if needed. (This can be useful for checking soil pH, cation-exchange capacity base saturation, and levels of nutrient elements such as potassium, phosphorus, and sulfur. Sampling may not be needed every season.)

4. If needed, you can also apply potassium sulfate fertilizer, either broadcast or shanked along side of the grapevines at rates ranging from 200 to 500 pounds per acre.

5. Cover crops are seeded. Many growers till every other row and seed it with mixes containing both legumes and small grains. The intervening rows may either be seeded with annual no-till cover crops such as subterranean clover, rose clover, bur medic, Blando brome, or Zorro fescue, or they may be left to reseed on their own with resident vegetation.

December, January, and February

1. Begin pruning. Most growers chop the prunings and let the residue break down gradually in the soil. Over the years, this contributes slowly to soil carbon.

2. If needed, apply boron sprays to the soil beneath the vines. Only apply boron sprays if petiole analysis indicates it is needed, as boron can be toxic if overapplied.

3. Zinc sulfate mixes can be dabbed onto vine cuts if there are indications of zinc deficiency.

4. Pruning weights are taken from selected vines marked for evaluation of shoot-to-fruit ratios. (These vines are marked, and will be harvested for crop load data.)

March, April, and May

1. No-till annual legume cover crops are mowed in early March with the mower set high to cut competing weeds.

2. Watch annual plow-down cover crops. At bloom time, you will mow them and incorporate them into the soil if nitrogen is needed in the vineyard. The timing can range from mid-April to early May, depending on the year. Mowing and incorporation should be done while the soil is still moist. If the year is dry and you have overhead sprinklers, you might consider applying 1 to 3 inches of water to ensure that the soil is moistened so that soil microorganisms can decompose the cover crop following tillage.

3. Carefully apply pre-bloom foliar sprays with zinc and boron. Usually these are mixed with wettable sulfur sprays. If you suspect that you have serious micronutrient problems, leave a small area unsprayed for petiole sampling.

4. Sample petioles for tissue testing and send them to a lab for analysis. If the petioles have been sprayed with micronutrients, rinse them off before you send them in for analysis.

June, July, and Early August

1. Potassium sulfate, if needed (based on petiole tests), can be applied beneath drip emitters any time during the summer. You can apply the material to the soil surface or place it into the soil by digging a small hole and dropping the fertilizer into it.

2. You can also place concentrated organic fertilizer or compost directly beneath the drip emitters. You get best results when you dig small holes, place the material in the holes, and cover it with soil. This should be done early in the summer, right after bloom, for best results.

3. Materials that can be fertigated can be applied with irrigations. This might include potassium sulfate, hydrolyzed animal proteins, and fish emulsion.

4. In late June or early July, evaluate the canopy for color and vigor. For most cultivars, shoots should be between 3 and 4 feet long with 15 to 20 leaves. Make note of the vines' appearance.

5. At veraison (usually mid-July to early August), evaluate vines again. Are the shoot tips dormant? Is the wood of the shoots starting to mature? If there is still considerable green growth, you may be overfertilizing or overwatering your vines. This is also the time to evaluate the vines for potassium deficiency (common) or phosphorus deficiency (rare).

Mid-August to Mid-October

1. Harvest test vines for canopy management: crop load vines, 10 in each block. Make sure that you have the vines individually tagged and identified.

2. Following delivery, ask the winery for a copy of any juice or wine sample data so that you can check to see if there were any problems with fermentation or quality that might be related to vine nutrition. When time allows, discuss your fruit quality with the winemaker, if possible.

3. When the vineyard has been picked, apply postharvest irrigation if you have sufficient water. If you use overhead sprinklers, apply 2 to 3 inches of precipitation, the ideal amount to start self-reseeding annual cover crops.

4. Start to prepare the seedbed for cover crops if you are using species that require tillage.

REFERENCES/RESOURCES

Anamosa, P. 2005. Management of high-magnesium soils for viticultural production. In Proceedings of the Soil Environment and Vine Mineral Nutrition Seminar, San Diego, California, June 29–30, 2004. Davis: American Society for Enology and Viticulture. Pp. 113–120.

Baver, L. 1956. Soil physics. New York: Wiley & Sons. Pp. 155–157.

Brady, N. C., and R. R. Weil. 2002. The nature and property of soils, 13th ed. Upper Saddle River, New Jersey: Prentice Hall.

Brinton, B., and A. York. 2003a. Sustainable composting in the vineyard. Part I: Basics of the process. Practical Winery and Vineyard 25(3).

———. 2003b. Sustainable composting in the vineyard. Part II. Practical Winery and Vineyard 25(4).

Cass, A. 2005. Effects of soil physical characteristics on mineral nutrient availability, movement, and uptake. In Proceedings of the Soil Environment and Vine Mineral Nutrition Seminar, San Diego, California, June 29–30, 2004. Davis: American Society for Enology and Viticulture. Pp. 3–11.

Chapman, P. 1985. Conservation farming. Soil Conservation Service of New South Wales, Australia. Pp. 12–20.

Gardiner, D. T., and R. W. Miller. 2004. Soils in our environment, 10th ed. Upper Saddle River, New Jersey: Pearson Prentice Hall.

Ingels, C., R. Bugg, G. McGourty, P. Christensen, and K. Scow. 1998. Cover Cropping in vineyards. Oakland: University of California, Division of Agriculture and Natural Resources, Publication 3338.

Koepf, H., R. Shouldice, and W. Goldstein. 1989. The biodynamic farm: Agriculture in the service of the Eatch and humanity. Hudson, New York: Anthroposophic Press.

Logan, W. B. 1996. Dirt: The ecstatic skin of the earth. New York: Berkeley Publishing. Pp. 7–17 and 45–54.

McGourty, G., and J. Reganold. 2005. Managing vineyard soil organic matter with cover crops. In Proceedings of the Soil Environment and Vine Mineral Nutrition Seminar, San Diego, California, June 29–30, 2004. Davis: American Society for Enology and Viticulture. Pp. 145–151.

Morlat, R., and A. Jacquet. 2003. Grapevine root system and soil characteristics in a vineyard maintained long-term with or without inter-row sward. American Journal of Enology and Viticulture 54(1).

Plaster, E. J. 2003. Soil science and management, 4th ed. Clifton Park, New York: Delmar Learning.

Reeve, J. R. 2003. Effects of biodynamic preparations of soil, winegrape, and compost quality on a California vineyard. M.S. Thesis. Washington State University, Department of Crop and Soil Sciences.

Reganold, J. 1988. Comparison of soil properties as influenced by organic and conventional farming systems. American Journal of Alternative Agriculture 3(4):144–155.

PHOTO · TOM LIDEN

Nutrients and Fertilizers for Organically Grown Grapevines

GLENN T. McGOURTY AND PETE CHRISTENSEN

See Tables 5.1 and 5.2 for information on grapevine requirements and deficiency levels for most nutrients described in this chapter.

NITROGEN (N)

Importance

Many vineyards in California need nitrogen (N) supplementation, and this is especially true in the San Joaquin Valley, where vines support larger canopies and crops. Vineyards in the state's North and Central Coast regions require less nitrogen because they have smaller canopies and crop loads. Typically, growers make relatively small annual applications of nitrogen, when compared to common application layers for other horticultural crops. Experience has shown that about 30 pounds nitrogen per acre per year is adequate for most vineyards, and this is consistent with measured harvest removals of nitrogen by grapes. For example, in one case 6.8 tons of grapes removed about 20 lb nitrogen per acre (2.94 pounds per ton of grapes).

Vines do not always show obvious symptoms of deficiency. The foliage appears pale green when a vine is nitrogen deficient. Usually, canopy growth and fruit yields will decline before obvious visual symptoms appear. Excessive nitrogen can cause rank growth, predisposing the vines to insect and disease attack, fruit shatter, and eventually reduced fruit set and poor winegrape quality, and possibly delayed fruit maturity.

Soil Sources

Nitrogen occurs in native soils as the end product of decomposition of the dead plant and animal tissues that make up a soil's organic matter. Unlike other nutrients, nitrogen is not derived from soil minerals.

Nitrogen can also be obtained directly from the atmosphere: rhizobium bacteria, which form nodules on leguminous plants such as alfalfa, clovers, vetches, peas, and beans, use a process known as nitrogen fixation. This process converts atmospheric nitrogen, which is not available to plants, into organic nitrogen in the soil, which eventually does become available for plant use. A legume cover crop can easily fix sufficient nitrogen to supply a vineyard's needs.

Rainwater, irrigation water, and well water can also be additional sources of nitrogen. Well water in some parts of California has nitrogen concentrations so high that the need for additional nitrogen fertilizer applications is eliminated when it is used to irrigate the crops. It is a good idea to periodically check your irrigation source for nitrogen, so you will know exactly how much nitrogen it can supply to the vineyard.

Role

Plants use nitrogen to build proteins that make up protoplasm, the basic living part of plant cells. Nitrogen is an important constituent of amino acids, lecithins, and chlorophyll. Nitrogen deficiency slows plant growth and causes carbohydrate reserves to increase. Excess nitrogen contributes to excessive crop growth and reduced carbohydrate accumulation.

Uptake and Utilization

Grapevines absorb most of their nitrogen in the nitrate form and move it readily from the roots to the leaves. Once there, it metabolizes into amino acids, proteins, and other nitrogen-containing compounds. Conversion from nitrates to these compounds is dependent on temperature and plant enzymes. Nitrate reductase converts NO_3^- to NH_4^+, and its activity varies from one variety to another.

Table 5.1. Summary of grapevine nutrient deficiencies and treatments

Element	Seasonal appearance of deficiency symptoms	Shoot position of deficiency symptoms	Common deficiency?	Appropriate for fertigation treatment?	Easily corrected?
Boron (B)	early	apical	yes	yes	yes
Zinc (Zn)	early	apical: mid-stem	somewhat	somewhat	somewhat
Iron (Fe)	early	apical: entire	somewhat	yes	somewhat
Manganese (Mn)	mid	basal	no	no	no
Phosphorus (P)	mid	basal	no	yes	somewhat
Potassium (K)	mid to late	mid	yes	yes	yes
Magnesium (Mg)	mid to late	basal to mid	yes, in San Joaquin Valley	yes	yes
Nitrogen (N)	not diagnostic		yes	yes	yes

Varieties that are high in nitrate reductase content are quite efficient at converting nitrates and generally have low nitrate content in their tissue. Others may be less efficient and may have higher nitrate content in their tissue. Cool weather and low light conditions can also affect the conversion process and in some years may cause high nitrate levels in the plant. There are no universal standards that apply to all varieties for nitrogen content in the vine tissue.

Diagnosing Nitrogen Deficiency

Plants that are low in nitrogen do not always show visual symptoms unless the deficiency is severe. Pale green to yellowish foliage, poor shoot growth, and low vigor are symptoms associated with nitrogen deficiency. Midsummer is a good time to evaluate vines for overall vigor and color. A root system that is impaired by insects, diseases, or poor physical conditions in the soil may be the actual cause of the symptoms, so you need to consider this possibility when diagnosing a deficiency. Fruit from these vines may have ripening problems and may also ferment poorly in the winery.

Diagnosing Excess Nitrogen

Plants with excess nitrogen will often have a dark green color, vigorous shoots, long internodes, and somewhat flattened stems. Because they have been shaded by excess foliage, many of the canes are green when they go into dormancy, making them more prone to freeze damage in cold areas. Often, buds are less fertile due to shading, and the more vigorous shoots may not be the best choices for fruiting wood during pruning. Vines with high nitrogen content have been observed to be more susceptible to leafhopper infestations, powdery mildew, and bunch rot.

Under excessively high soil nitrogen conditions, plants may be subject to temporary toxicity in the spring when cool weather follows periods of warm weather. The petioles of affected leaves have high nitrate levels, and mostly the older leaves are affected. They have a deep green color and, often, glossy patches on the upper leaf surface. Protein compounds (amino acids) exude from pores (hydathodes) at the leaf edges, leaving a white, saltlike deposit. In severe cases, tissue at the leaf edges (or sometimes entire leaves) will die. Close examination may reveal a water-soaked appearance in some of the blade tissue before it turns brown. Symptoms are usually short lived and occur before bloom. Warm weather encourages rapid nitrogen uptake, but the nitrogen compounds are not adequately assimilated or utilized in the new growth. Toxic levels of nitrates or intermediate compounds then concentrate during the cool weather that follows. Affected vines typically start to grow normally once warm weather returns. This syndrome is sometimes known as "spring fever." Leaves look as if they are potassium deficient, and they usually are low in potassium, but not to the degree that would be considered "deficient." Early

Table 5.2. Critical plant tissue values and application of nutritional standards for practical use in vineyards*

Nutrient	Concentration range					
	Deficient	Marginal	Critical (deficiency)	Adequate	High	Toxic or excessive
N (%)				0.8-1/10		
NO₃-N (mg/kg)	<340	340-499		500-1,200	≥1,200	
P (%)	<0.2	0.2-0.24		0.25-0.50	>0.50	
K (%)	<1.0	1.0-1.7		1.8-3.0		
Ca (%)				1.2-2.5		
Mg (%)	<0.3	0.3-0.39		>0.4		
Na (%)						>0.5
Cl (%)						>1.0-1.5
Fe (mg/kg)				>30		
Cu (mg/kg)	<3	3-5		6-11		
Zn (mg/kg)	<15	16-25		>26		
Mn (mg/kg)	<20	20-29		30-60		>500
B (mg/kg)	<25	26-34		35-70	71-100	>100

*Standards for grapevines (from Robinson et al. 1997): **Growth stage:** when the majority of the vines are flowering. **Plant part:** petiole (leaf stock) of a basal leaf opposite a leaf cluster; one petiole from each of 100 vines throughout a planting. **How established:** California standards of Cook (1966) and Christensen et al. (1978) modified following survey work in South Africa (Robinson and McCarthy 1985) and field trials in Victoria (Treeby and Nagarajah, CSIRO unpublished data; further work is in process to relate petiole standards to vine vigor) and in Western Australia (WA) (Goldspink, unpublished data, Western Australia Department of Agriculture). Weir and Cresswell (1993) present leaf blade standards for both flowering and veraison sampling times. *Adapted from Robinson 2005.*

bunch stem necrosis is the necrosis and drying of flowers and the cluster/stem structure just before and during bloom. This disorder is also associated with an accumulation of toxic nitrogen compounds.

Young vines 1 to 3 years old sometimes exhibit leaf burn symptoms when they are excessively fertilized. One indicative symptom is marginal browning of the leaves. In severe cases, all of the leaves turn brown and show amino acid exudate and water-soaked tissue. If nitrogen is the cause, plant tissue testing should show a high nitrogen content. To alleviate symptoms, discontinue fertilization and apply heavy irrigations to leach some of the fertilizer through the root zone. Excess nitrogen is very unusual in vineyards grown under organic conditions, though, since most organic fertilizers are not highly soluble. Even so, you should be careful when applying compost or concentrated nitrogen fertilizers to young vines.

Nitrogen Fertilization and Organic Winegrowing

Vineyards need about 30 pounds of nitrogen per year. This can be supplied easily by compost and green manure cover crops, and even by the irrigation water in some areas. For organic winegrowers, the use of cover crops and compost is the most common approach to supplying nitrogen needs (more information in chapter 6, Cover Cropping Systems for Organically Farmed Vineyards). Most growers apply about 1 ton of compost per acre per year and plant leguminous nitrogen-fixing cover crops on at least 25 percent of the vineyard floor. The plant response to nitrogen is fairly rapid (within 1 month after application) and should certainly be evident within the current growing season if you make applications early in the season. Nitrogen from compost may take more time to reach the plants since mineralization is not always fast and depends on the presences of low C:N ratios, soil moisture, and warm temperatures.

Other nitrogen sources include concentrated organic fertilizers such as feather meal, blood meal, and fish protein concentrates. These materials contain about 10 to 13 percent nitrogen by weight as well as small amounts of phosphorus and potassium (less than 1%). The release of nitrogen from these materials is controlled by microbial biological activity, and soil moisture and warmth are needed for nitrogen to mineralize. Mineralization takes about 8 to 12 weeks, making these materials behave very much like slow-release fertilizers.

Some commercial blended fertilizers available also have higher concentrations of phosphorus and potassium. These materials are used primarily in new plantings or during transition to organic, as an organic soil fertility program is being developed. Often, growers apply these products beneath drip emitters or bury them in the soil beneath the emitters.

Concentrated organic fertilizers can be injected through a vineyard's drip system if you have the right equipment and material. Spray-dried animal proteins (which have a very small particle size) and some fish emulsion fertilizers can be used for organic "fertigation." The injection equipment should be upstream of the drip system's filtration unit to prevent the accidental clogging of emitters. Always flush the drip system after a fertigation application.

POTASSIUM (K)

Importance

Grapevines use potassium (K) in quantities comparable to their nitrogen use. Potassium deficiency is common in the North Coast vineyards and less so in the San Joaquin Valley growing areas. The availability of potassium depends in part on the type of minerals from which the soil has been formed. Soils formed from granitic rocks, as in the San Joaquin Valley, tend to have adequate potassium. However, some young granitic soils include vermiculite, a mineral that fixes added potassium in mineral interlayers, making it unavailable to plants and increasing the need to supplement the soil with potassium fertilizers. Soils formed from ultramafic, basaltic, and serpentine rocks high in magnesium, as in the Coast Range and Sierra Foothills, can often be borderline or deficient in potassium. Potassium tends to be stable in the soil and does not readily leach. Clay minerals can also reduce the soil's cation-exchange capacity (CEC), as potassium ions may be held tightly between alumina-silicate layers. Sandy soils are also likely to show potassium deficiencies due to low CEC values and low content of potassium-releasing minerals.

Sources in the Soil

Most potassium in soils is derived from weathered minerals, especially feldspars and micas. These minerals are only slightly soluble and are found as large particles in the soil. As the particles weather, the potassium is solubilized and becomes available to plants as positively charged ions (cations). Most of these ions are held by the cation-exchange complex associated with clay colloids or organic matter in the soil. While there may be a fairly large amount of potassium in a given soil, most of it will be insoluble and only a small percentage will be available to plants.

Potassium fertilizers applied to the soil may be fixed by clay minerals or held in the cation-exchange complex and slowly released over time. Clays and clay loams are most likely to fix the most potassium. Sands and sandy loams fix less potassium but are more prone to potassium leaching and so more likely to have a deficiency. In any case, potassium leaches slowly and does not move readily into the soil profile.

The potassium cycle is fairly simple. Plant-available potassium in the soil solution and on cation-exchange sites on clay and organic matter is taken up by the grapevine roots. This available potassium is replenished by the weathering of potassium-containing soil minerals, which are plentiful in many California soils. Where this natural supply is not adequate for plant potassium needs, you need to turn to fertilizers, manures, and composts. The vine's potassium uptake is aided by good soil structure and a large root system.

Role and Utilization

The precise role of potassium in grapevine physiology is not completely understood. Better understood is the effect of potassium deficiencies on vine growth and crop yields. Plants need potassium for the formation of sugars and starches, for the synthesis of proteins, and for cell division. Potassium also neutralizes organic acids, regulates the activity of other mineral nutrients in plants, activates certain enzymes, and helps adjust water relationships. It is also involved in a grapevine's cold hardiness and its formation of carbohydrates, even though potassium is not usually found as part of organic compounds. About 1 to 4 percent of plant tissue by dry weight is made up of potassium.

The demand for potassium is highest during fruit ripening in mid- to late summer, when large amounts of the nutrient accumulate in fruit. Temporary deficiencies may be associated with excessively large crops.

Diagnosing Potassium Deficiency

Usually, a noticeable response to potassium fertilizer will only occur when deficiency symptoms are present. There is no reason to add additional fertilizer to vines that have an adequate amount of potassium. Fertilizer trials conducted in vineyards with low (but not deficient) levels of potassium typically show no yield or growth response. Soils with a low level of potassium in the soil solution can become deficient. This can be a fine-textured soil with high magnesium content or low potassium in the cation-exchange complex and in sandy soils. Symptoms are likely to occur at veraison and to continue through ripening. Potassium is mobile in the plant, moving to the ripening fruit from foliage, which turns yellow or brown. In most crop years, each ton of winegrapes will remove between 5 and 8 pounds of potassium from the soil. The vines' potassium uptake is also strongly influenced by soil moisture conditions and root density and health. Water stress can contribute to potassium deficiency, particularly during fruit ripening, when potassium demands are highest. If you have a vineyard with marginal potassium levels in the soil, you will want to monitor regulated deficit irrigation very closely. Phylloxera and nematode damage to root systems may also reduce potassium uptake.

Deficiency Symptoms

Leaf symptoms begin to show in early summer on leaves in the middle portion of the shoots. Marginal yellowing begins, and as the season progresses the yellowing spreads into areas between the main veins, leaving a central island of green extending around the main veins. In red-fruited varieties, the yellowed areas may bronze or redden. In all varieties, marginal burning and curling usually follow. Symptoms may develop in leaves above and below the middle of the stem right up until harvest time.

Severe potassium deficiency reduces shoot growth, a symptom that may be visible before bloom. Fruit development often is poor, with small, tight clusters and unevenly colored small berries. Fruit maturity may be delayed. Damaged root systems from a variety of causes, including high water tables in wet years, phylloxera, and soilborne plant diseases, may also cause vines to show signs of potassium deficiency. It is important to check root health whenever you observe nutrient deficiencies in a vineyard.

Potassium Fertilization in Organic Winegrowing

In developing a potassium fertility program, you may need to address other factors such as soil structure, high magnesium, and other problems that may require the application of gypsum or composts and cover cropping to improve the soil environment for roots and facilitate the exchange of potassium in the cation-exchange complex. Since potassium is not readily available from organic fertilizer sources, most growers apply potassium assuming that the application will take time to mineralize and that it may take more than 3 years before they can expect to see a response. It is also important to remember that potassium fertilizer will elicit no yield response unless the vines are deficient. In fact, excess potassium in fruit will raise the pH in the resulting wine and lower its acidity, which affects taste and wine stability.

Compost can provide an adequate amount of potassium for a vineyard, particularly if the compost is made from recycled grape pomace. A ton of compost can provide between 20 and 60 lb of potassium, depending on the source material for the compost. Annual compost applications made in the fall can provide adequate potassium for vines, particularly if the application is made directly beneath the vines.

An increase in soil organic matter (SOM) has two positive effects on potassium availability. First, potassium that is held on exchange sites on SOM is less likely to become fixed, whereas the soil on its own does have the capacity to fix potassium. Second, soils with a higher level of SOM are likely to be better structured and it is likely to be easier for roots to come into contact with potassium-exchange sites.

PHOSPHORUS (P)

Importance

Grapevine phosphorus (P) deficiencies in California are rare. Where they do occur, they are most likely to be observed in soils with unusual characteristics, such as soils with very low or very high pH or soils that originated from volcanic ash parent materials. Phosphorus deficiency has become more common, though, as new vineyards have been planted in cooler or wetter areas with acidic soils. In soils with pH values below 5.0, the solubility of aluminum increases. The aluminum is then taken up by plant roots, where it interferes with the uptake and utilization of phosphorus, calcium (Ca), and magnesium (Mg). As a grower, you may find it necessary to amend such soils with a liming material to increase the pH level to one more favorable for vine growth.

Sources in the Soil and Uptake

The chemistry of phosphorus in the soil is complex. Phosphorus usually is bonded with calcium, iron, and aluminum, as well as carbon in the humus fraction of the soil. Its solubility is low, and it mineralizes only slowly in the soil. In low- and high-pH soil conditions, phosphorus becomes available extremely slowly, and under such conditions a deficiency might be possible. The mineralization of phosphorus from organic matter is affected by pH and other environmental conditions that also affect soil microbiology. Growers sometimes notice a deficiency after soil fumigation. This is because soil mycorrhyzal fungi are involved in phosphorus uptake. Following fumigation, commercial nurseries apply phosphorus in anticipation of this problem. Some growers address this issue by re-inoculating their soils with a mix of beneficial fungi following fumigation.

Of the fertilizer trials that have been conducted in the San Joaquin Valley, many have failed to show a response to phosphorus fertilizer, even with application rates as high as 2,000 pounds of treble super phosphate per acre.

Phosphorus Deficiency Symptoms

Visual symptoms of phosphorus deficiency are seldom seen in California. In many different plant species besides grapes, symptoms include stunting and reddening of the foliage. These symptoms are most commonly seen early in the year when the soil is cold and new roots are growing slowly. Symptoms also become evident near veraison, when reddening occurs in the older foliage. In soils with low pH (below 5.0) or high pH (above 8), phosphorus might become deficient due to precipitation with other elements or interference by other elements in root uptake. Soil amendments should be added to bring the pH into a more favorable range; otherwise, phosphorus applications are unlikely to improve vine nutrition. If you suspect that your vines have a phosphorus deficiency, it is probably best for you to consult with a farm advisor or crop consultant: the conditions that cause phosphorus deficiency are complex, and correction may require several steps.

Phosphorus Fertilization in Organic Vineyards

In most cases it will not be necessary to apply phosphorus to your vineyard, as phosphorus is rarely deficient in grapevines under California conditions. However, cover crops may benefit from additional phosphorus, particularly in high rainfall regions. Nearly all legumes require phosphorus to grow well. Since phosphorus tends to be deficient in North Coast soils, it should be considered as part of the soil fertility program for vineyards in that area.

Humus particles adsorb phosphate ions to their surface. These ions are more readily available to growing plants than is phosphorus that is precipitated as insoluble compounds (often with calcium). Often, soils with high levels of humus have more available phosphorus than low-humus soils. Building soil organic matter has many benefits if phosphorus is deficient, especially in more alkaline soils. Since many composts (and especially those made from manures) contain significant amounts of phosphorus, they can help build up available phosphorus in the soil. Composts are between 0.5 to 2 percent phosphorus. When applied at 1 ton per acre, then, compost puts between 5 and 20 pounds of usable phosphorus into the soil. Poultry manure is also high in phosphorus and would be another good choice, especially in acid soils, due to the manure's high pH. USDA National Organic Program (NOP) rules require that poultry manure be composted before application. (A note of caution: The high nitrogen content of composted poultry manure can cause excessive vine vigor if too much is applied.)

If vineyard soils are acidic, there may also be some benefit to applying rock phosphorus. Rock phosphorus is between 14 and 35 percent phosphorus and is applied at fairly high rates (up to 500 lb per acre) because it takes many years for the material to mineralize. Bone meal is another option if the soil has a low pH. Phosphate content varies depending on the source of the bone meal and how it was processed, but most bone meals contain about 30 percent phosphorus and are applied at around 100 to 300 pounds per acre. Bone meal is about four times as expensive as rock phosphate, but it may mineralize more quickly. Remember that these applications are likely to directly benefit your cover crops more than your grapevines, since grapevines are less sensitive to low phosphorus levels.

MAGNESIUM (Mg)

Importance

Magnesium (Mg) deficiency in grapevines is rare in California, where it occurs mostly in vineyards that are planted on sandy soils having a low cation-exchange capacity (CEC). In the San Joaquin Valley, deficiencies have also been noted in sodium-affected soils, soils that have been heavily fertilized with manure, and vineyards with a long history of drip irrigation and gypsum fertigation. Some rootstocks and varieties (such as rootstock 44-53 and the varieties Grenache, Thompson Seedless, Barbera, Malbec, and Gamay) may show signs of deficiency while others show none. Young vines are the most susceptible to magnesium deficiency, and symptoms often disappear as the vines' root systems expand into subsoils that have a higher magnesium content. Deficiencies sometimes show up in topsoil fill areas after land leveling, especially if the fill soils are sandy. In other areas, magnesium sometimes is reported to be deficient after heavy applications of potassium fertilizer. While not common in California, there have been reports of magnesium deficiency in vineyards with a long history of drip irrigation and fertigation with potassium (50 to 100 lb per acre) in the San Joaquin Valley.

Soil Availability and Uptake

California soils generally have high levels of available magnesium, and in some instances those levels are excessive. The mineral sources are dolomite, biotite, and serpentinite. Plants absorb magnesium as the Mg^{++} ion from the cation-exchange complex where, like potassium, it is attached to negatively charged clay particles and organic matter. The roots extract magnesium either from the soil solution or from contact with soil colloids.

Role and Utilization

Magnesium is a constituent of chlorophyll and is essential for photosynthesis. It also activates many enzymes required for plant growth. Magnesium is highly mobile within plants and can translocate readily from older to younger tissue when there is a deficiency.

Symptoms and Diagnosis of Magnesium Deficiency

A vine is showing signs of deficiency when its basal leaves become chlorotic (yellow), normally after midseason. The chlorosis will progress up the shoots to the younger leaves. Chlorosis begins on the leaf margins and then moves inward between the primary and secondary veins. Some border of normal, green color remains along the veins, and may become almost creamy white. The chlorosis is often followed by leaf margin burn. Many red-fruited varietals also show a reddish border inside the burned perimeter of tissue. Symptoms become more severe with large crop loads and are most obvious as the fruit become ripe. At least 30 percent of the vine canopy must show symptoms before yield reductions are likely to occur.

Symptoms and Diagnosis of Excess Magnesium

Soils that are derived from serpentine rocks often contain very large amounts of magnesium. These soils occur primarily in the Coast Range and the Sierra Foothills. Many of these soils are poorly structured and low in potassium, and they may have toxic amounts of both nickel and manganese. (Nickel is toxic when present at over 50 ppm.) You need to evaluate such areas carefully before you opt to establish a vineyard there. Soil tests evaluating the CEC and base saturation can help you determine whether there are potential problems with excess magnesium. If the base saturation of the soil contains 60 percent or more magnesium, the soil probably has too much magnesium. A Ca:Mg ratio of less than 1:1 is also potentially troublesome and most likely will require soil amendment with gypsum to improve soil physical conditions and potassium availability. If the site is sloping, it may also be physically very unstable and prone to slippage when wet. Vines often grow poorly in these locations, and potassium often is deficient. Corrective actions involve large applications of gypsum (preplant if possible) and large quantities of potassium sulfate applied to the soil, as well as cover cropping, compost applications, and other practices that help to improve soil structure.

Fertilization and Responses to Magnesium

Magnesium deficiencies are not common in California, and in economic terms they may not be important enough to treat. If symptoms become apparent by midsummer, you may want to treat the vineyard. Suggested rates are 2 to 4 pounds of Epsom salts per vine as a banded application. It may take several seasons for the fertilizer to have its effect. Magnesium applied by drip fertigation and foliar sprays is also effective.

Magnesium Fertilization in Organic Vineyards

Although this is rarely necessary in California, you can apply magnesium to organic vineyards in the form of mined minerals. Dolomitic lime contains about 14 percent magnesium and would be quite useful for correcting a deficiency if you also need to raise the soil's pH. This is a very cost-effective treatment. Application rates vary from 1 to 3 tons per acre and are determined in part by soil pH. If there already is sufficient calcium in the soil and the soil's pH is close to neutral, you might want to consider using a different fertilizer source.

Magnesium sulfate (from mined sources) can easily be supplied to vines via fertigation. If you use a drip system, do not use any other fertilizers that contain phosphorus or calcium, as that might result in precipitation and plugging. Magnesium precipitates can also form if carbonates are present in the water, so make sure to flush the system well after injections. You can also hand apply magnesium to the vines beneath drippers, as $MgSO^4$ is quite soluble. Foliar applications of magnesium sulfate also appear to work quickly. One or two applications of $MgSO^4$ made before bloom and after fruit set at 4 pounds per 100 gallons of water per acre is considered an effective rate.

Sulfate of potash magnesia (20% S, 20% K_2O, 10% Mg), also known as Sul-Po-Mag, is another naturally occurring mineral form of magnesium that can be used in organic vineyards. It is applied at the rate of 300 to 500 pounds per acre. Note, though, that this is an expensive treatment. Compost is another source of readily exchangeable magnesium, ranging between 10 to 30 pounds per ton of finished material. Its effectiveness as a magnesium fertilizer is unknown.

SULFUR (S)

Importance

Sulfur (S) deficiencies in grapevines are virtually unknown. However, leguminous cover crops may suffer from sulfur deficiencies that are not easily identified. Symptoms on cover crops include young leaves with a light green or yellowish color, small and spindly plants, poor growth, and delayed maturity.

Source in the Soil and Uptake

Sulfur is found in nature as elemental sulfur, sulfide, and sulfates, and in organic molecules with carbon and nitrogen. The original terrestrial sources of sulfur were probably sulfides of metals contained in plutonic rocks. During weathering, these minerals decomposed and the sulfide was oxidized and released as sulfate. Precipitation into soluble and insoluble sulfate salts occurred in drier climates, and these materials were utilized by living organisms or reduced to sulfides or elemental sulfur under anaerobic conditions. Most of the sulfur present in soil is combined with organic matter, part of the soil solution (as sulfates), or adsorbed onto the soil's anion-exchange complex (AEC).

Uptake of sulfur by plants is in the form of the sulfate ion (SO_4^-). Sulfur dioxide may also be absorbed from the air through leaves. Because elemental sulfur is commonly applied as a fungicide, grapevines usually have an ample supply. If you are using sulfur dust at rates above 30 pounds per acre per year, you probably are applying enough to ensure that there is sufficient sulfur for all of the plants in the vineyard, including cover crops. Air pollution containing various sulfur pollutants can provide sulfur to foliage, and to the soil when it rains. If your vineyard is located in a region with poor air quality, it is highly unlikely that sulfur will be deficient in the soil. In some areas on California's North Coast, sulfur levels in the soil can be quite low as a result of the clean air, clean rain water, high rainfall, leaching, and soil parent material that is low in sulfur. The SO_4^{--} ion can leach readily from the soil profile or precipitate to form gypsum in lower-rainfall areas.

Sulfur is found in organic matter, usually with nitrogen in a N:S ratio of 10:1 or 10:1.5. The presence of adequate sulfur in the soil may affect the decomposition of carbon-containing materials. When the C:S or N:S ratio is too wide, subsequent plant growth for annual crops is poor due to the immobilization of sulfur, and when the C:S or N:S ratio is too narrow, immobilization of nitrogen may cause poor plant growth. These conditions rarely affect grapevines, but your cover crops may be affected if you are not using compost.

When elemental sulfur is applied to the soil, it is not soluble. It is slowly solubilized as biological activity converts it to SO_4^-, which is usable by plants. The soil bacterium *Thiobacillus* oxidizes sulfur into the SO_4^- ion. The rate of conversion depends on the particle size of the sulfur, the temperature, and the soil moisture level. Conditions that favor the decomposition of organic matter also favor the oxidation of sulfur into SO_4^-.

Role and Utilization

Sulfur is an important component of three amino acids (cystine, methionine, and cysteine), which are important in protein synthesis. These compounds are essential for nodule formation on legume roots, and this is important if you are growing leguminous cover crops in your vineyard.

Fertilization in Organic Vineyards

Plants require sulfur in about the same quantities as they require phosphorus. Compost contains about 10 to 15 pounds of sulfur per ton, a quantity sufficient to provide all of the sulfur that grapevines need in a vineyard. Sulfur in compost takes the form of SO_4^- or of sulfur in organic molecules, which is converted to SO_4^- over time. Cover crops normally respond quickly to compost applications. If you do not use compost, gypsum and elemental sulfur are two alternative sulfur fertilizers available to you. Gypsum is slightly soluble and, in high-rainfall areas, more than 70 percent of its sulfur content can leach beyond the root zone. Apply elemental sulfur in mixed particle sizes so that mineralization will occur over time, with the smaller particles oxidizing first because they have a greater surface area, and thus more contact per unit of weight with soil bacteria. Remember that sulfur is an acidifying fertilizer and may lower the soil's pH, particularly when applied to soils that are already acidic. An application of 40 pounds of sulfur dust per acre, applied to the soil, will be adequate for most cover

crops if sulfur is deficient. For areas prone to sulfur deficiencies, an application of 100 to 200 pounds per acre of sulfur in mixed sizes will last 3 to 8 years.

GRAPEVINES AND MICRONUTRIENTS

ZINC (ZN)

Importance

Zinc (Zn) deficiency is common in many California vineyards, and many growers routinely fertilize to correct this problem. It is most common in the San Joaquin Valley and the Central Coast regions, and somewhat rare on the North Coast. The deficiency is sometimes called "little leaf disease," in reference to common foliar symptoms. Zinc deficiencies have also been noted in soils with basic (high-pH) soil reaction, after heavy applications of compost and manure, and after soil fumigation with methyl bromide.

Sources in the Soil

Zinc is found in minute quantities in most soils. Sandy soils have the lowest levels overall. The product of weathered minerals, zinc is adsorbed by clay and organic matter and held in an exchangeable condition. Zinc levels are highest in the upper horizons of the soil, especially where organic matter is being cycled from crop residues.

Zinc becomes less available when the soil pH is greater than 6. At pH 9, it is virtually unavailable. In soils high in organic matter, calcium, magnesium, and phosphorus, zinc may be unavailable to plants.

Zinc deficiency occurs most frequently in vineyards planted on sandy soils, sites that were formerly corrals or poultry houses, or places where land leveling has caused heavy cuts to bring former sub-horizons to the surface. Some vigorous rootstocks such as Dogridge, Salt Creek, Freedom, and Couderc 1613 are prone to zinc deficiency if planted to these areas.

The application of high rates of nitrogen fertilizers may accentuate a zinc deficiency when the nitrogen stimulates total vine growth and so increases the vines' zinc needs, which then exceed the available supply. Rapidly growing young vines sometimes exhibit temporary deficiency symptoms, but they frequently grow out of the problem as their root systems expand and come into contact with a larger soil surface area.

Zinc deficiency is very common following fumigation and planting. Like phosphorus, the uptake of zinc is probably dependent on mycorrhyzal fungi, which help roots take up some nutrients.

Role and Utilization

Zinc is critical for the formation of auxins, chemicals that enable the elongation of internodes and in the formation of chloroplasts and starch in the vine. In grapes, zinc is also essential for normal leaf development, shoot elongation, pollen development, and the set of fully developed berries.

Symptoms and Diagnosis of Zinc Deficiency

Even before visible symptoms become evident on grapevine foliage, zinc deficiencies may cause reduced fruit yields. Foliar symptoms vary, depending on the severity of zinc deficiency. These symptoms include mottling, which appears in early summer when lateral shoot growth is well developed. The new growth on both the primary and secondary shoots has smaller, somewhat distorted leaves that show a chlorotic pattern that sets off the darker, greener color of the veins. Even the smaller veinlets retain a uniform-width border of green unless the deficiency is quite severe.

Severely affected leaves have undeveloped basal lobes, and only a shallow sinus with little indentation where the petiole joins the leaf. "Little leaf" is a good description of the overall appearance of affected vines. Shoots are stunted, with closely spaced, small, distorted leaves when deficiency is severe.

Fruit set can suffer significantly when zinc deficiency is severe. Clusters tend to be straggly, with fewer berries than normal vines. Berry size varies greatly and may include small "shot berries," which tend to remain green. Merlot is very prone to poor fruit set in the North Coast; foliar zinc applications applied pre-bloom help to alleviate this problem. Uneven ripening is also noted in some cases. Some varietals like Muscat of Alexandria exhibit fruit symptoms without showing foliage symptoms. Other varietals show both simultaneously.

Zinc deficiency symptoms are usually obvious to the eye, but you may want to confirm the diagnosis with tissue tests. Sometimes growers confuse zinc deficiency symptoms with the symptoms of other micronutrient deficiencies or of fanleaf virus. If your vineyard has no history of zinc deficiency and suddenly symptoms develop, tissue testing is particularly advisable. Soil testing is not considered helpful: it is difficult to correlate the zinc level in the soil with that in the vine. Rootstocks differ in their ability to forage for zinc. Some may find adequate amounts of zinc even when the overall soil content is low.

Fertilization and Response to Zinc

Zinc deficiencies are fairly easy to fix, and growers have several methods they can follow. Pre-plant incorporation of zinc sulfate is the recommendation of many consultants, but with this method most of the zinc becomes fixed and unavailable near the soil surface. Pre-bloom foliar sprays with zinc sulfate can be very effective. Soil applications can be effective on sandy soils. Soil applications are mostly limited to the treatment of chronic or severe deficiencies, which are difficult to correct with foliar sprays. Soil applications are more effective on sandy soils. To be effective, a soil application must place the material deep in the soil, near the vines.

Zinc Fertilization in Organic Vineyards

There are several sources of zinc that can be used in an organically certified vineyard. Most are applied as pre-bloom sprays and contain either zinc sulfate or chelated zinc. Many are tank-mixed with wettable sulfur applications and applied pre-bloom. Neutral zinc products containing 50 to 52 percent zinc are the most cost effective. Zinc sulfate (36% Zn) or zinc chelate can be injected into drip irrigation systems. From 50 to 100 pounds of zinc sulfate per acre or 3 to 6 pounds chelated zinc (6.5% solution) per acre is recommended for a mature vineyard with a moderate to severe deficiency. Growers should carefully consider the costs of this practice versus the cost of foliar sprays. The advantages of drip application are labor savings and a longer-lived response.

BORON (B)

Importance

Boron (B) is an important micronutrient for winegrapes. Grapevines need boron in very small quantities, and the range between deficiency and toxicity is surprisingly small: about 0.4 ppm of boron in the soil solution is sufficient, and 1 ppm and above is probably toxic.

Boron deficiencies occur in many parts of California. In the San Joaquin Valley, boron may be deficient in sandy soils that formed as alluvium from granitic parent material, mostly on the eastern side of the valley. Irrigation water sources that are low in boron tend to leach this micronutrient from the soil profile. In northern California, high rainfall tends to leach boron from the soil profile so that deficiencies sometimes occur. Toxicity is also a problem, when springs and seepage are naturally high in boron or when high levels of boron are introduced from irrigation water, whether from wells or surface water. There is no effective treatment to remove excess boron from irrigation sources. If multiple water sources are available, you may be able to dilute high-boron water with water from another source that is lower in boron content.

Sources

Boron occurs naturally in the form of borosilicate minerals, which are resistant to weathering and release boron slowly. Much of the available boron is held by the organic and clay fractions of the soil in the anion-exchange complex. Boron is less leachable than are other neutral or negatively charged plant nutrients.

Boron also occurs naturally in well water and surface water, particularly in the Coast Range and the west side of the San Joaquin Valley. It is common in areas with geothermal activity.

Uptake, Utilization, and Role

Boron is taken up by plants as borate and is involved in the differentiation of new cells. When boron is deficient, cells may still continue to divide, but their structural parts will not be completely or properly formed. Boron also regulates carbohydrate metabolism. During bloom, low boron levels limit pollen germination and normal pollen tube growth. This can result in poor pollination and fruit set.

Boron does not move from older tissues to newer, and young leaves and shoots are the first to show deficiency symptoms. A continuous supply of boron must be available during the growing season for normal growth to occur.

Diagnosing Boron Deficiency

Boron deficiency symptoms can vary widely and are easily confused with the symptoms of other disorders. Two categories of deficiency are recognized: a temporary, early spring deficiency, and an early to mid-summer deficiency.

Temporary, early spring deficiency begins to appear after budbreak. Stunted, distorted shoots characterize the early growth, but most of the shoots begin to elongate normally again as the season progresses. Affected shoots are dwarfed because their internodes fail to elongate and may grow in a zig-zag manner. Numerous lateral shoots will grow from the stunted shoots, giving the plant a bushy appearance. The growing tip may die on severely affected shoots. Affected shoots are unfruitful or have undeveloped clusters. Often, these symptoms occur after a particularly dry fall and early winter. This type of boron deficiency is considered to be temporary. Low soil moisture conditions are known to restrict boron uptake severely and can contribute to a deficiency.

For example, severely affected Grenache leaves are somewhat fan-shaped and may show an interveinal chlorosis; the serrations around the leaf edges will be irregular, and the veins will be more prominent than on normal leaves. Affected Chenin Blanc leaves have a wide, fan-shaped appearance and prominent veins. Barbera and Chardonnay leaves have a more rounded but misshapen appearance.

Early to midsummer deficiency occurs more consistently from year to year than does early spring deficiency. Symptoms show during or just after flowering and are most pronounced on berry set and growth. Low boron levels during the rapid growth period before and after bloom cause the problem. Leaf tip dieback is common, but it is followed by normal growth, which can make stunted shoots hard to identify later in the season. The vine will have adequate boron levels later as growth slows into summer unless the vine is stressed for water.

Fruit symptoms are the easiest to recognize. Severely affected vines may have no crop at all. Some clusters may appear to burn off or dry up around bloom time, leaving only cluster stems, sometimes with occasional berries. Many clusters may set numerous small, seedless berries that persist and ripen. These clusters may be full or straggly, and they may also include some normal-sized berries as well as shot berries. Shot berries are distinct and uniform in size and shape. Unlike the normally more oval or elongated berries of most varieties, shot berries are very round to somewhat flattened, resembling in profile a very small tomato or pumpkin. Shot berries caused by low boron levels should not be confused with those caused by zinc deficiency. Clusters from zinc-deficient vines include shot berries that have the normal shape for their variety, and most of those berries remain hard and green. They are also more varied in size than the shot berries caused by boron deficiency. Occasionally, clusters on boron-deficient vines appear at bloom time to have set normally, but will shatter severely in midsummer.

Boron Fertilization in Organic Vineyards

Boron deficiencies are easily corrected by a variety of methods, but they only require relatively small amounts of fertilizer, and only those small amounts are recommended as safe. It is very important that you stay within recommended application rates, calibrate your equipment properly, and mix solutions to the right concentration to avoid vine toxicity.

A standard vineyard rate of 1 pound actual boron per acre per year is used with most application methods. Hand or spray broadcast applications usually apply 4 pounds boron per acre, but with the expectation that the treatment will last approximately 4 years. This method assumes that rainfall will move the boron into the soil and requires more attention to re-treatment timing, which will vary depending on your vineyard site and rainfall conditions. Application through drip irrigation is also effective and safe at the 1 pound per acre rate, provided that the vines are fully mature. However, you may have to modify application rates in successive years to compensate for potential increased boron concentrations in the vine. Foliar sprays of boron can be used as a temporary or emergency treatment, although some growers use this as their routine approach. Usually, applications are made pre-bloom, or postharvest if the canopy is still functioning. After one or more years of treatment, there should be enough residual boron in the vineyard soil to provide a year-round supply to the vines. After that point, you can apply boron at any time of year to achieve the 1 pound boron per acre per year recommendation for maintenance. A spring or summer foliar application should not exceed ½ pound boron per acre in any given application in order to avoid phytotoxicity symptoms. Vine foliage is more tolerant of boron in the fall, when it can safely receive up to 1 pound boron per acre in a single application.

Fall applications are also more effective at preventing symptoms that would otherwise occur early in the following season. Boron that is incorporated into the vines' dormant tissues will prevent subsequent flower and fruit set symptoms.

Most soluble boron products are derived from mined sodium borate. One example is Solubor, which is 20.5 percent boron (5 lb of Solubor is equivalent to about 1 lb of actual boron). Compost also contains small amounts of boron, usually less than 1 pound per ton, but these levels are probably too low to be useful for correcting deficiencies. Always monitor boron treatments with tissue analysis to make sure you do not exceed the vines' narrow range of tolerance.

IRON (Fe)

Importance

Iron (Fe) deficiency is considered the third most common micronutrient deficiency in California grapevines, behind zinc and boron. Fortunately, most iron deficiencies are temporary and isolated. Iron deficiencies are often persistent and severe, however, in high-lime content soils and with susceptible varieties. Symptoms are most likely to occur when vines are planted on soils that are high in limestone or that have either high or low pH. Iron deficiency may also be noted on vines with weakened root systems or those that have been over-cropped in previous seasons or where the available iron in the soil is limited.

Soil Availability and Uptake

Iron occurs in soils as oxides, hydroxides, and phosphates, as well as in the lattice structure of clay minerals and some silicates. As an element, iron is abundant in soil. Under varying soil conditions, small amounts of iron are released during the weathering of minerals and are absorbed by roots in the ionic form or as complex organic salts. In its ionic form, iron is highly reactive, easily oxidized, and present in the soil solution in very small amounts.

Deficiency symptoms occur primarily where soil conditions limit root uptake and vine utilization of iron rather than where total iron is lacking in the soil. In California, high-pH soils that are affected by high phosphate content, high lime (calcium carbonate) content, or high alkali (saline-sodic) content are most likely to show iron deficiencies. In these soils, a condition know as lime-induced iron chlorosis occurs, in which iron is immobilized or inactivated by the high levels of carbonates or lime in the soil.

Vines growing in fine-textured soils, especially when the soils are poorly drained and cold, are more likely to show iron deficiency symptoms if saline-sodic or high-lime content conditions are also present. Iron deficiency is often noted in the spring when the weather is cool and wet. Vines often "grow out" of the problem as the weather warms. However, vines can also show symptoms during periods of rapid growth and warm temperatures. Temporary iron deficiencies have also been noted when vineyards are heavily irrigated in the spring.

Utilization and Role

Iron is transported in plants as the ferrous ion (Fe^{++}) and becomes immobile as it is complexed into various organic compounds. Iron is not transported readily from one tissue to another in the plant, and new growth is more likely to show deficiency symptoms.

Iron functions in the activation of several enzyme systems. A shortage of usable iron also impairs chlorophyll production, resulting in the characteristic chlorosis.

Diagnosing Iron Deficiency

Foliar symptoms first appear as interveinal yellowing on shoot tip foliage. As the leaves expand, the blade appears pale yellow with clearly defined green veins (as opposed to zinc deficiency, where the green pattern is more diffuse). As deficiencies worsen, the leaves may become ivory or white, and even brown and necrotic. Growth is reduced on severely affected shoots, tips are stunted, and the fruit cluster rachis is likely to be chlorotic. Fruit set is likely to be poor on affected shoots. As grapevines recover from temporary iron deficiencies, new growth develops with a normal green color. Color improvement is delayed, though, on mature leaves that were affected earlier in the season. Soil and tissue analysis have not been found useful in diagnosis because soil and plant tissue iron levels do not necessarily correspond to the occurrence of deficiency.

Correcting Iron Deficiency Symptoms

Iron deficiency is considered to be one of the most difficult nutritional problems to correct in plants, and grapevines are no exception. Foliar-applied iron solutions have only limited success since grapevine leaves do not absorb or translocate these materials very well. Soil-applied iron sulfate and other inorganic compounds do not move well through the soil , and under high lime or saline-sodic soil conditions they are rapidly inactivated by oxidation and complexing. Iron chelates work better, but they are quite expensive and their effects are not always long lived. More successful approaches involve applying soil amendments such as sulfur, gypsum, or sulfuric acid to improve the soil's rooting environment. Soil amendment with organic matter also increases the availability of iron. Recent work in San Joaquin Valley vineyards has shown that ferrous sulfate applied in the spring with fertigation is as effective as chelated iron materials are, and it is less expensive.

If an area is chronically iron deficient, consider using lime-tolerant rootstocks. Three of the rootstocks commonly used in California—5BB, 420A, and 140 Ru—have medium-high tolerance. Medium-tolerance rootstocks include 1103P, 110R, 5C, and St. George. In Europe, 41B and Fercal are used in the most severe cases, but little is known about how well they will perform in California.

Iron Fertilization in Organic Vineyards

Iron chelates and ferrous sulfate dissolve readily in water and can be applied easily through a microirrigation system. This is a relatively expensive treatment, though, and requires springtime fertigation every year. Balancing the expense of the treatment, though, is the probably greater long-term expense that results when you leave deficiencies untreated. Inorganic salts of iron are not usually effective when applied to the soil surface under high-pH conditions.

The improvement of soil conditions with soil amendments to bring the soil pH into a more favorable range is probably the best long-term strategy for dealing with this problem. The most permanent solution is to plant a more tolerant rootstock in problem areas.

MANGANESE (Mn)

Importance

Manganese (Mn) deficiency is not common in California vineyards, and when it does occur it is rarely severe. It occurs most often in sandy soils. Manganese toxicity is sometimes a problem in serpentine soils. High manganese levels may interfere with the availability and uptake of iron by plants.

Sources in the Soil

Soil manganese originates from the decomposition of ferro-manganesian rocks. Manganese occurs in the soil solution as an oxide as well as an ion that is absorbed and exchanged in soil colloids and organic matter. Soil pH influences availability, with increases as the soil becomes more acidic (lower pH).

Manganese often becomes available to plants when the soil becomes temporarily anaerobic after irrigation or a rain. The temporary lack of oxygen in the soil causes manganese oxides to dissolve slowly. The small amount of manganese that does dissolve is often available for several days after the anaerobic conditions begin. Consequently, manganese deficiencies are not common unless the soil is naturally low in manganese-containing compounds.

Utilization and Role

Plants take up manganese in the ionic form (Ma^{++}). It is a relatively immobile nutrient within the plant. Manganese serves as an activator for enzymes in growth processes. It assists in chlorophyll formation. Leaf chlorosis is an early deficiency symptom.

Diagnosing Manganese Deficiency

Symptoms can appear on severely deficient vines 2 or 3 weeks after bloom. A mild to moderate deficiency will not become apparent until mid- to late summer. The symptoms begin on basal leaves as a chlorosis between the veins. Increasing chlorosis develops between the primary and secondary veins; the veinlets tend to retain a green border. A somewhat distinct herringbone chlorosis pattern can eventually develop on manganese-deficient leaves.

These symptoms are easily distinguished from those of zinc, iron, and magnesium deficiencies. Zinc deficiency symptoms appear first on newer growth and include some leaf malformation. Iron deficiency also appears on newer growth and causes a much finer network of green veins in the yellowing leaf tissue. Like manganese deficiency, magnesium deficiency chlorosis first appears on the basal leaves, but it is more extensive than manganese deficiency between the primary and secondary veins, developing into more complete yellowish bands (more of a "Christmas tree" pattern) that lack the herringbone pattern. You should turn to laboratory analysis of petiole samples from affected leaves and normal leaves for a final diagnosis.

Correction of Symptoms

Manganese deficiencies in California are considered rare and inconsequential. Most symptoms occur on older leaves late in the season and so have only a minimal effect on vine physiology and fruit ripening. Few trials have been done on manganese deficiencies in California. Foliar sprays of either manganese sulfate or chelated manganese have been used with some success by growers. Manganese sulfate (28.5% Mn) application at 1 to 2 pounds per acre is the most economical treatment.

Manganese Fertilization in Organic Vineyards

It is unlikely that it would be necessary for you to apply manganese to a vineyard in California. If you do find that treatment is desirable, foliar treatment with manganese sulfate is probably the easiest and most efficient way to fix a deficiency. The treatment of deficiencies with soil-applied manganese fertilizers is not well understood. Manganese is not bound well by chelates and tends to drop off chelation molecules when applied to soil. Often, the chelation molecules pick up iron instead, and carry it into the plant tissues. This results in an imbalance that might actually accentuate the manganese deficiency. Because the soil tends to fix manganese, neither fertigation nor broadcast applications of manganese sulfate are likely to alleviate deficiency symptoms.

REFERENCES/RESOURCES

Brady, N. C., and R. R. Weil. 2002. The nature and property of soils, 13th ed. Upper Saddle River, New Jersey: Prentice Hall. Pp. 498–542.

Brinton, B., and A. York. 2003a. Sustainable composting in the vineyard. Part I: Basics of the process. Practical Winery and Vineyard 25(3):16–27.

———. 2003b. Sustainable composting in the vineyard. Part II. Practical Winery and Vineyard 25(4):50–50.

Burt, C., K. O'Connor, and T. Ruehr. 1998. Fertigation. San Luis Obispo, California: Irrigation Training and Research Center, California Polytechnic State University.

Campbell, A., and D. Fey. 2003. Soil management and grapevine nutrition. In Oregon viticulture, E. W. Hellman, Ed. Corvallis: Oregon State University Press. Pp. 143–162.

Christensen, L. P., and D. Smart, Eds. 2005. Proceedings of the Soil Environment and Vine Mineral Nutrition Seminar, San Diego, California, June 29–30, 2004. Davis: American Society for Enology and Viticulture.

Gardiner, D. T., and R. W. Miller. 2004. Soils in our environment, 10th ed. Upper Saddle River, New Jersey: Pearson Prentice Hall. Pp. 126–165.

Hirschfelt, D. 1998. Soil fertility and vine nutrition. In Cover cropping in vineyards, C. Ingels et al., eds. Oakland: University of California, Division of Agriculture and Natural Resources, Publication 3338. Pp. 61–68.

Parnes, R. 1990. Fertile soil: A grower's guide to organic and inorganic fertilizers. Davis, California: agAccess.

Reganold, J. 1988. Comparison of soil properties as influenced by organic and conventional farming systems. American Journal of Alternative Agriculture 3(4):144–155.

Scow, K., and M. Werner. 1998. Soil ecology. In Cover cropping in vineyards, C. Ingels et al., eds. Oakland: University of California, Division of Agriculture and Natural Resources, Publication 3338. Pp. 69–79.

Shanks, L. W., D. More, and C. Epifanio Sanders. 1998. Soil erosion. In Cover cropping in vineyards, C. Ingels et al., eds. Oakland: University of California, Division of Agriculture and Natural Resources, Publication 3338. Pp. 80–85.

Soil Improvement Committee. 1998. Western fertilizer handbook, 2nd horticultural ed. Danville, Illinois: Interstate Publishers, Inc.

Tisdale, S., and W. Nelson. 1975. Soil fertility and fertilizers, 3rd ed. New York: McMillan Publishing Company.

White, R. 2003. Soils for fine wines. Oxford: Oxford University Press.

Cover Cropping Systems for Organically Farmed Vineyards

GLENN T. McGOURTY

Since 1990, growers and researchers have made substantial advances with cover cropping practices for the vineyard. When organic and sustainable winegrowing began in the late 1980s, growers adopted many of the cover crop species already in use in agronomic crop farming systems. While these were well suited for stimulating soil biological activity and soil improvement, they had not been selected specifically as ideal species for the vineyard. Many species were simply too vigorous: they grew into trellises, produced too much biomass to manage easily, and produced too much nitrogen for the soil in most vineyards. Often, these cover crops were very energy intensive to farm, requiring multiple tillage operations, extra fertilizer applications, and seeding operations beyond those used for standard vineyard floor management in that era.

More recent research and grower experiences have led growers to choose cover crops that are better suited to the architecture of the vineyard as well as the farming systems that growers find appropriate for their vineyards. Protection from soil erosion and building soil structure, organic matter, and overall soil quality are also obvious goals when you plant cover crops.

Cover cropping is now widely practiced in vineyards around the state of California and is considered an important practice for many reasons. The University of California publication *Cover Cropping in Vineyards* (Ingels et al. 1998) is an excellent reference on this subject. In this chapter, we will examine how organic winegrowers select cover cropping systems and how they grow and manage them.

BENEFITS OF COVER CROPS IN ORGANIC VINEYARDS

Cover crops help winegrowers manage their soils in several ways:

- *Protecting the soil from erosion.* Cover crop foliage reduces the velocity of raindrops before they hit the soil surface and in this way reduces the amount of soil splashing. This prevents the slaking of soil aggregates and the sealing of the soil surface (a sealed soil surface increases runoff, and with it, soil erosion). The roots of cover crops bind soil particles together, improving soil's structure and water penetration while preventing erosion of the soil particles.

- *Regulating vine growth.* You can use cover crops to invigorate vines (by using nitrogen-fixing legumes to augment soil nitrogen) or to reduce vine vigor (by letting nonlegumes compete with the vines for nutrients and water).

- *Improving soil fertility.* Leguminous cover crops can fix atmospheric nitrogen if they are inoculated with the proper strain of Rhizobia bacteria, which will live symbiotically in their roots. When the leguminous plant tissue decays, the nitrogen becomes available in the soil in amounts that may range from a low of 20 pounds per acre (from self-reseeding annual clovers) to a high of nearly 300 pounds per acre (from berseem clover). Most winegrape vineyards require only 30 lb of nitrogen per acre per year, so leguminous cover crops can easily supply these needs. In fact, you can easily get an overabundance of nitrogen if you use leguminous cover crops in a vineyard planted on fertile soil. Besides increasing soil nitrogen, decomposed cover crops increase the soil's cation-exchange capacity. This increases the soil's ability to hold and exchange nutrients. In addition, nutrients are often chelated into organic complexes and are more readily exchanged from these substrates than from inorganic clay minerals.

- *Improving soil structure and water-holding capacity.* Initially, cover crops (especially grasses) help to aggregate soils as their fine roots penetrate the soil profile. The fine roots also help prevent soil erosion. Cover crops that have large taproots help to create macropores in the soil. When the plants die, they leave a void where their roots have decomposed. These macropores greatly increase the movement of air and water into the soil profile. Soil organisms that use the decomposing cover crops as a food source create waxes and other sticky substances that bind the fine particles into aggregates, lowering the soil's bulk density and improving its tilth. As the soil's organic matter content increases, so does its ability to hold water. Physical improvement of the soil is important because in organic winegrowing, a large root system is very desirable. Since soil nutrient concentrations may be lower in an organic vineyard than in a conventional vineyard and since many organic winegrowers in coastal areas prefer not to irrigate, a root system that forages through a larger area of the soil profile is more likely to supply the water and nutrients that the vine needs. Organic growers have experienced measurable improvement in wine quality from fruit grown in vineyards where they use organic soil management techniques that ultimately improve soil physical conditions.

- *Enhancing biological diversity in the root zone.* Organic matter is a food source for macro- and microorganisms. Many of these organisms help recycle the cover crops into the soil and improve soil physical qualities in the process. Particularly noteworthy are increases in earthworm populations; the presence of earthworms is a good indicator of soil health and improved physical conditions. Increased biological activity occurs in the soil after you incorporate organic matter from cover crops. The soil organisms can

reduce damage from root pathogens and inhibit the growth and development of those and other pathogens. As an example, many phylloxera-infested vineyards planted on AXR1 rootstock are still producing crops at economically viable levels in organic vineyards more than 14 years after the first insects were found. Most conventionally farmed vineyards fail to produce an economically viable crop just 6 years after phylloxera are found. The ultimate cause of the failure is secondary infection from fungi that invade phylloxera-damaged roots. Researchers believe that competition from soil microbial populations keeps those fungal pathogens from causing damage.

- *Providing habitat for beneficial generalist predator and parasitoid insects and arachnids (spiders and mites).* Since pest management strategies in organic winegrowing emphasize a reliance on nature and the use of "soft" chemicals, it is important that beneficial arthropods be abundant and in close proximity to the vineyard so they can control harmful insects and mites. Cover crops can provide habitat and food for beneficial insects at different stages of their life cycle. They also provide habitat for their prey, such as aphids, mites, caterpillars, and other creatures. Research entomologists have a difficult time understanding the complex dynamics of pest and prey relationships in the cover crop and their effects on the grapevine canopy. Regardless, there are numerous accounts by growers who observed reductions in leafhopper and mite problems when they planted cover crops and stopped applying conventional insecticides. As an example, back in the 1980s when vineyard floors were tilled, one vineyard management company treated nearly 1,200 vineyard acres in Mendocino County's Russian River Valley annually with dimethoate for leafhopper control. Today, many of those acres are cover cropped or managed by mowing the resident vegetation. The use of insecticides for leafhopper control has ceased. Based on vineyard monitoring, growers and PCAs give credit to the biological control of leafhoppers and mites by generalist predators and parasitoids for eliminating the need to spray. These beneficial insects and spiders are readily found both in the vines and in the cover crops.

- *Providing a firm footing for harvest and cultural operations.* When no-till, sod-forming cover crops are planted, vineyard operations during wet weather become more feasible as the cover crops provide firmer footing for equipment. This can enable harvest, pruning, and spraying during inclement weather.

- *Improving air and water quality.* Water quality law enforcement is on the increase, and vineyard water runoff needs to be free of silt and nutrients. Cover crops help reduce silt by preventing erosion. Until it is mineralized, the nitrogen formed by legumes is less mobile than the soluble nitrogen in conventional fertilizers. The cover crops assimilate free nutrients in the soil and stabilize them during periods of high rainfall: for example, nitrogen assimilation by *Phacelia* spp., mustards, and annual grains, and phosphate assimilation in compost by legumes during winter months. During dry periods of the year, cover crops help reduce dust by covering the soil and protecting surface particles from wind, thus improving air quality. This also helps reduce mite infestations, which thrive under dusty conditions.

COVER CROPPING FARMING SYSTEMS

Your choice of a cover cropping farming system will depend on the relative vigor of your site; the availability of water in the soil; your viticultural objectives (increasing or decreasing vegetative growth); and your pest management objectives for insect, mite, and weed control. Here are a few different approaches:

Annually tilled and seeded. Most growers using the annually tilled and seeded system choose it as a way to conserve moisture in their vineyards. The cover crop is planted in the fall, allowed to grow until a point in the spring when the ground can still be easily cultivated, and then mowed and tilled into the soil. This last operation is often timed to occur when the cover crop is flowering, since cover crops decompose easily at this stage. This system is best suited for relatively flat vineyards in which soil erosion does not pose a serious potential hazard. In upland areas that are prone to soil erosion and where water is not available for irrigation of either the vines or the cover crop, you should use straw mulching or compost "overs" (coarse particles between 1 and 2 inches across) to minimize the loss of soil from the vineyard while you wait for the cover crop to start growing in the fall. In vineyards that use drip irrigation, many growers mow these annual cover crops in the spring and let the residues lie on the surface. The growers then till in the fall just before reseeding.

Cover crop species typically used in this system include annual small grains (barley, oats, triticale), winter peas, common vetch, bell beans, daikon radish, Persian clover, and other annuals that grow well during the cool months of the year. This farming system is tillage intensive, and soil is exposed to sunlight during the summer. Too much tillage will cause a loss of soil structure and organic matter. Even so, many growers utilizing this system grow very high-quality fruit without irrigation or concentrated fertilizers and their vineyards give yields comparable to those of conventional vineyards. Many people like the looks of a cultivated vineyard, and often this is the method of choice near expensive and attractive winery facilities, particularly in the Napa Valley.

No-till vineyard floor management with annual cover crop species. In this system, the vineyards are first tilled and then seeded with species that will annually reseed themselves. Thereafter, the grower mows the vineyards in spring and early summer. Tillage is only done beneath the vines. Subterranean clovers, rose clovers, crimson clover, red clover, berseem clover, bur medic, bolansa clover, and Persian clover are all well suited for this farming system. You can also use certain grasses, including Blando brome and Zorro fescue.

Another no-till approach is to plant annual cover crops that are not self-reseeding, such as oats, barley, peas, and vetch, using a no-till drill. This is a useful approach where tillage could cause erosion and where it is desirable to keep tillage to a minimum. Usually, seeding is done in the fall just before rains begin. The standing cover crop is then mowed and left to lie on the soil surface.

No-till vineyard floor management with perennial species. Perennial species are most commonly used in vineyards that are growing on fertile sites. Many of the perennial grasses are very competitive with grapevine roots and so have a devigorating effect on the vineyard. This may be desirable if the vineyard is seriously out of vegetative balance, such as when a vine's fruit-to-shoot weight ratio is less than 4:1. There is a range of cover crops that vary from slightly competitive to very competitive. The fine fescues (hard fescue, creeping red fescue, and sheep fescue) are the least competitive, grow very short, and survive well. Turf selections of perennial ryegrass and tall fescue are intermediate in their competitiveness. They are also fairly low statured, so they require mowing only once or twice per year. Finally, pasture selections of perennial ryegrass, tall fescue, and orchard grass are the most competitive and can have a tremendous impact on the vigor of a vineyard. You should plant them on only the most vigorous and deep-soiled sites.

These perennial grasses may also be used in parts of the vineyard that are prone to erosion or areas where you want to reduce dust. Seasonal waterways, vineyard roads that are not heavily trafficked, turn-around areas, staging areas, and other places where the soil needs to be protected are potential sites for perennial grasses.

Research conducted in pastures has shown that when you include perennial legumes in a sward of grasses they will supply nitrogen for the grasses. Unfortunately, they may also attract rodents such as voles and gophers that can also damage grapevines. Despite this potential problem, many growers do include white clover, strawberry clover, alsike clover, and birdsfoot trefoil in their perennial cover crop mixes. Besides providing nitrogen for the grasses, these species provide habitat for generalist predator and parasitoid insects. Some growers have also had success planting perennial grasses alone and then, after two or three seasons, planting annual legumes into the sward. If the annual legumes and perennial grasses are initially planted together, the legumes will shade the perennials out and the sward is likely to have a poor stand of grasses.

Finally, there are also California native grasses that can be used as cover crops. Favorites include pine bluegrass, Mokelumne fescue, and Molate red fescue (less-competitive species), and California brome, meadow barley, and blue wild rye (more-competitive choices). Seeds for these grasses are expensive, and in some cases they do not compete with weeds as well as other introduced pasture grass species that are used as cover crops. It is also important to let these grasses flower late in the spring so they will be able to accumulate carbohydrates in their root systems, improving their persistence and competitiveness with weeds.

Using both tilled and no-tilled farming systems. Some growers use different farming systems in alternate rows to moderate the vines' vigor, incorporate compost, or provide diverse habitat, or for aesthetic reasons. One system commonly employed uses a no-till system of self-reseeding annuals for 3 years in alternate rows, with annually planted cover crops (which are plowed down every year) in the other rows. After 3 years, the growers switch the no-till/self-reseeding rows to annually planted/plowed-down, and vice-versa. Perennial species are also used in this way in alternate rows. In most cases, this approach is used on more vigorous sites that are not prone to soil erosion.

Cover crop rotations. Over time, a cover crop can develop pests and pathogens to the extent that it becomes difficult to reseed the same species year after year. That is one of the reasons that growers plant seed mixes. The negative effects of planting the same species every year seem less pronounced when you use a mixture of diverse species.

Sometimes growers will use completely different species from year to year. For instance, they will plant mustards or radishes one year, legumes the next, and legumes with annual grains the year after that. Other growers mix all three and plant them together, since the combination of species that thrive in the sward varies with each season's unique growing conditions.

CULTURAL PRACTICES FOR COVER CROPS

Seed source. Cover crop seeds should be purchased from dealers who sell high-quality seeds that have been tested for viability and are free of weed species. Under organic certification laws, growers are obliged to attempt to source organically grown seeds. Unfortunately, this is nearly impossible for many of the small-seeded cover crops, which are not even grown in the United States. In your certification records you must document in writing that you attempted to purchase organic seeds, but with that documentation you can plant conventionally grown cover crop seeds in your organically certified vineyards.

Seeding. Most cover crop species grow best when planted in a well-prepared seedbed with adequate fertility. Usually, this requires at least two diskings, harrowing, and firming the soil with a ring roller or cultipacker prior to seeding. If the ground is somewhat compacted, you may have to shallow-rip the planting area to a depth of about 12 inches using a tool bar and shanks, especially in areas with wheel traffic in the vineyard middles.

For seeding, you can use any of a number of implements. For small areas, you can use hand broadcast spreaders ("belly grinders"). Tractor-mounted broadcast spreaders can be used for larger areas, but they are not very precise. Seed drills are the best choice if you are planting expensive seeds and require accurate placement.

Most seed drills use two soil-cutting blades (coulters) set at acute angles to each other. These cut a slit in the soil, and seeds are metered from a box above them, falling through tubes that open between the two coulters. Small wheels behind the coulters pack the soil firmly after the seeds are deposited. Another setup uses a ring roller attached to the seed drill to firm the soil after seeding.

Slit seeders are used for no-till seeding. These utilize a device similar to a rotary tiller, except that the flat cutting blades radiate straight out from the hub rather than being bent at right angles as is usual for a rototiller. The seed box is mounted above the tiller, and seeds are directed into the slits made by the rotating blades. The soil is then packed by a ring roller mounted on the seeder. This seeder works best for cover crop species that have considerable seedling vigor.

Time of seeding and irrigation. Growers usually plant cover crops in the fall and rely on fall rains to start germination. In cooler areas where the growing season is shorter, many vineyards are equipped with overhead sprinkler systems for frost protection. In these areas, it is very helpful to seed early and irrigate the vineyard with 1 inch of water in late September to mid-October to start the germination process. The germination of small-seeded cover crops and perennial species definitely benefits from early seeding and irrigation. If rains don't come immediately, you may need to apply additional water.

Perennial species can also be seeded in the spring along with warm-season summer cover crops, assuming that overhead irrigation is available to assist germination. In California's interior valleys, these crops can be seeded from February through March. In the North Coast, planting should be delayed until late April or early May. In many respects, this is not the best time to seed perennials. A fall seeding is more likely to provide them with the moist soil conditions they need in order to develop an extensive root system. Perennial species will need to be mowed to reduce competition from annual summer weeds. Properly timed irrigation will benefit both annual and perennial cover crops.

Fertilizer. Cover crops need specific nutrients to grow well. Many organic growers use compost, which will in most cases adequately provide what the cover crops need. Compost made from a mixture of animal manure and grape pomace (50:50 mix) normally has enough nitrogen, phosphorus, and potassium (N-P-K) to get the cover crops off to a good start. Rates vary, but most growers will start with 1 or 2 tons per acre applied annually. Legumes respond well to rock phosphorus applied one season and physically incorporated into the soil, followed by liming the next season. Base your application rate and timing decisions on soil tests to ensure that you apply the proper amounts of materials. Vineyards in some high-rainfall areas require popcorn sulfur applications, specifically for the benefit of legume cover crops. Rangeland soil fertility studies of subterranean clover pastures in the 1960s showed a clear response to sulfur fertilization. High amounts of rain cause sulfur deficiency in many North Coast soils. The rainwater in that area is free of sulfur because there is little air pollution there or over the Pacific Ocean, where the rain clouds form.

Springtime mowing. Many cover crop species benefit from at least one springtime mowing, as it can eliminate shading from faster-growing weed species and promote tillering or expansion of the plants' crowns. For low-statured cover crops, this should be done just before they transition from the winter dormant/basal rosette stage into mature growth and flowering. In the North Coast, this usually occurs in early March, about the time that you would shred prunings, and the two operations are usually done with a single pass of the equipment.

Larger-statured annual cover crops (such as oats, bell beans, and mustards) are often clipped at vine budbreak, removing anything growing above 18 inches to lessen frost hazard for emerging vine growth. Tall cover crop swards are not desirable at this time, as they can impede air movement and increase the tendency of young shoots to freeze or develop Botrytis shoot tip rot.

Springtime tillage. If the vineyard is going to be disked, you will get the maximum addition of nitrogen from legumes if you incorporate them into the soil when they are blooming. Many growers will first shred the cover crop with a mower and then disk it in. This practice improves decomposition rates since smaller pieces of crop residue decompose more quickly. Timing is very important, as the soil must still be moist enough for you to easily till in the crop. Maximum nitrogen release occurs about 3 weeks after incorporation, assuming that the soil remains moist. Full incorporation of all residues may require multiple tillage operations, usually about 10 days apart. A final pass is often made with a ring roller to pack the soil firmly, making it easier to walk on and visually more attractive.

Late spring and summer mowing. Self-reseeding annual cover crops are mowed in the late spring and early summer after seed set to minimize dry residual growth that might be flammable and to mow down summer weeds. For perennial cover crops, you may need to do several mowings to keep the vegetation from growing too tall. If you cover crop with California native grasses, there are some advantages to letting the plants flower in June, and then mowing them. Even though large amounts of foliage can accumulate, this practice allows the grasses to accumulate more carbohydrates in their root systems so they will survive summer dormancy better than they would if continuously mowed.

CONCLUSION

Cover cropping is an important component in organic winegrowing and involves making numerous choices in which species to plant and which farming systems to use. Organic winegrowing practices do not exclude any cover crop choices that are also available to conventional winegrowing systems. The choice of a cover cropping system is very site specific and is best chosen by the growers to suit their style of farming, yield and quality objectives, and any other criteria that they consider important.

REFERENCES/RESOURCES

Ingels, C., R. Bugg, G. McGourty, and P. Christensen. 1998. Cover cropping in vineyards. In Cover cropping in vineyards, C. Ingels et al., eds. Oakland: University of California, Division of Agriculture and Natural Resources, Publication 3338.

McGourty, G. 1994. Cover crops for north coast vineyards. Practical Winery and Vineyard 15(2):8–15.

Miller, P. R., W. L. Graves, W. A. Williams, and B. A. Madson. 1989. Cover crops for California agriculture. Oakland: University of California, Division of Agriculture and Natural Resources, Publication 21471.

Sustainable Agriculture Network. 1998. Managing cover crops profitably. Handbook Series 3. Beltsville, Maryland: Sustainable Agriculture Research and Education Program of CSREES, U.S. Department of Agriculture, National Agricultural Library.

P A R T   3

Pest Management in Organic Vineyards

PHOTO: TOM LIDEN

7

Weed Management for Organic Vineyards

W. THOMAS LANINI, GLENN T. McGOURTY, AND L. ANN THRUPP

A weed can be defined, very simply, as a plant that is growing out of place. However, there is a group of plants, primarily nonnatives, that are considered to be weeds because they share certain other attributes:

- fast growth and high degree of competition with other plants for physical space, nutrients, light, and moisture

- prolific seed production

- high germination percentage and good seedling vigor

- short life cycle

- in most cases, herbaceous

We classify weeds based on their life cycle:

- *Winter annual weeds* germinate in the fall, overwinter as a basal rosette of leaves (broadleaf plants) or a tillering bunch (grasses), and then rapidly elongate and bloom during the spring (Table 7.1). As the weather warms by midsummer, they set seed and senesce. Most weeds in California

probably fall into this category, since they grow during the cool and wet parts of the year.

- *Summer annual weeds* germinate in the spring, grow actively during the warm times of year, and die either when soil moisture is depleted or when the weather turns cold in the fall (Table 7.1). These weeds are most common in places that are irrigated or where soil is moist during the summer, such as in riparian areas or along waterways.

- *Biennial weeds* germinate during one year and then flower and finish their growth cycle the next year (Table 7.2). Many spend the first year as a basal rosette of leaves, needing either cold weather or short day length to stimulate them to initiate growth of reproductive stalks, produce seed, and finish their life cycle.

- *Perennial weeds* survive for more than two seasons (Table 7.3). Typically, they grow vegetatively and produce seed and additional propagative structures in the warmer months and then go dormant when temperatures turn cold. Food reserves stored in the

Table 7.1. Some common winter and summer annual weeds found in California vineyards

Winter annuals	Summer annuals
Annual bluegrass (*Poa annua*)	Annual morning glories (*Ipomoea* spp.)
Common chickweed (*Stellaria media*)	Barnyardgrass (*Echinochloa crus-galli*)
Filarees (*Erodium* spp.)	Common purslane (*Portulaca oleracea*)
Groundsel (*Senecio vulgaris*)	Large crabgrass (*Digitaria sanguinalis*)
Mustards (*Brassica* spp.)	Spurges (*Chamaesyce* spp. *and Euphorbia* spp.)
Red maids (*Calandrinia ciliata*)	Yellow foxtail (*Setaria pumila*)
Wild radishes (*Raphanus* spp.)	
Yellow starthistle (*Centaurea solstitialis*)	

Table 7.2. Common biennial weeds found in California vineyards

Bull thistle *(Cirsium vulgare)*
Common mullein *(Verbascum thapsus)*
Poison hemlock *(Conium maculatum)*
Wild carrot *(Daucus carota)*

Table 7.3. Common perennial weeds found in California vineyards

Field bindweed *(Convolvulus arvensis)*
Nutsedges *(Cyperus* spp.)
Johnsongrass *(Sorghum halepense)*
Bermudagrass *(Cynodon dactylon)*

propagative structures are then used for the next season's growth. These weeds can be especially difficult to control if they are a species that is able to propagate both vegetatively and from seed. A classic example is Bermudagrass, which can grow from seed, stolons (aboveground lateral stems), and rhizomes (belowground lateral stems). Perennial weeds are particularly difficult to manage in irrigated crop areas or places with deep, fertile soils. Paradoxically, tillage can actually help to spread these weeds. Field bindweed, another perennial weed, also thrives in tilled vineyards: the plant regrows easily from underground rhizomes and can be very competitive with young vines.

Weeds are objectionable in vineyards because they can

- compete with vines (especially young ones) for space, nutrients, and moisture

- create habitat for damaging rodents such as pocket gophers and voles

- harbor insects or diseases that might damage vines directly or serve as vectors of grape pathogens

- be unsightly in the vineyard

Resident vegetation (often regarded as weeds) is sometimes considered a good cover crop because it represents a mix of adapted species and usually is easy to maintain. In some parts of the state, resident vegetation may include species that earlier growers planted in what were then pastures (and are now vineyards) and continue to reseed and grow year after year. These include subterranean clover, birdsfoot trefoil, bur medic, and others, and many

are actually recommended as cover crop species. Lime and compost applications to a vineyard often encourage them to grow. In Cover Cropping in Vineyards (Elmore, Donaldson, and Smith 1998), the authors state

> Plant life in the vineyard could be beneficial or harmful to the commercial production of grapes. The amount of vegetation and the time of year that it is present can cause various degrees of positive or negative impacts on grape production. Young vines, vines on shallow or weak soil, or areas that are commonly droughty are more susceptible to competition from other plants. However, when managed appropriately, plants may enhance vine growth by modifying soil fertility, improving water infiltration, and increasing organic matter.

The choice of appropriate methods for managing weeds and other vegetation in an organic vineyard is site specific and depends on the growth stage of the vineyard. There is no single correct approach: each vineyard's weed control program should be developed based on soils, weed species present, age of the vineyard, availability of water, potential for erosion, and aesthetic standards. Grapevines grow during the warm, dry part of the year. Winter annuals, if mowed or cultivated, do not compete too much with vines, especially if the grower uses drip irrigation in the summer. Some advantages of mowing weeds over cultivating them include

- The sod formed by untilled weeds can protect hillside vineyards from erosion and can reseed itself.

- Some mowed weeds provide habitat, pollen, and nectar for spiders, beneficial insect predators, and parasitoids, as well as habitat for their prey. They help to increase the number and diversity of insects in the farm ecosystem.

- During the growing season, the amount of dust in a mowed vineyard is lower than in a tilled vineyard. When you reduce dusty conditions, you also lower the incidence of spider mites.

- Mowed weeds provide firm footing during wet periods of the year when you have to enter the vineyard, whether on foot or in vehicles

GENERAL WEED MANAGEMENT ISSUES FOR ORGANIC GROWERS

In organic vineyards, objectives for weed control vary depending on the site, the soil, the grower's aesthetic preferences, costs, and the age of the vineyard. Typically, organic growers use the following management strategies:

- clean tillage during the growing season if the vineyard is young or if water for irrigation is limited or unavailable

- weed control beneath the vines by tillage, hand hoeing, flaming, or mowing, to reduce habitat for harmful rodents, reduce competition between weeds and grapevines, and keep the vineyard looking aesthetically pleasing

- mowing of vineyard row middles several times a year to reduce moisture competition and improve aesthetics

Reducing Weed Pressure

Weed pressure is the degree to which weeds propagate and inhibit crop growth. The more a grower is able to reduce weed reproduction (seeds and, for perennials, propagules) now, the less money the grower will have to spend on weed control in the future. Research indicates that without some form of weed control in a vineyard, yields and plant vigor will greatly suffer (Zabadal and Dittmer 2001). First-year vine shoot and leaf weights were reduced by more than 70 percent and second-year vines had more than an 85 percent reduc-

tion in leaf area when there was no weed control used under the vines, as compared to weed-free vineyards.

Management of the naturalized vegetation and weeds in an organic vineyard requires the use of several techniques and strategies if you want to achieve economically acceptable weed control, vine growth, and grape yield. Weeds can always be removed by hand, but the cost (both in terms of time and money) can be prohibitive.

Some growers of organic crops other than winegrapes make an extra effort to keep their farms free of weeds, mainly because they worry about fungal diseases that may be hosted by the weeds. A few weed species—Bermudagrass and horseweed, for instance—also host the glassy-winged sharpshooter (a vector for Pierce's disease), but this sort of disease-host relationship for weeds is not a major concern in most winegrowing regions of California. The weed-free approach is not common among organic winegrowers. Most growers who use it do not irrigate their vineyards, so weed elimination is a strategy for conserving moisture.

In contrast, many organic winegrowers are *not* focused on keeping the vineyard weed-free and bare. They recognize that low-growing weeds and resident vegetation can be considered acceptable or beneficial because they minimize soil erosion or add organic matter to the soil when they are cultivated or mowed. Moreover, many types of naturalized plants on the vineyard floor do not compete much with mature vines, provided they are mowed short and prevented from growing directly beneath the vine rows.

Some resident vegetation, however, may include perennial or annual weeds that are persistent and strongly competitive for nutrients and water and therefore need to be managed. Ideally, a grower should prevent any weed from producing mature seed because it can aggravate weed problems for years (Norris 1981). As an example, the seeds of common purslane (*Portulaca oleracea*) have been shown to remain viable for over 20 years in the soil, and black mustard seeds survive for over 40 years. The longevity of weed seeds, together with the large numbers of seeds produced by a single plant (e.g., 100,000 seeds per plant for large purslane or barnyardgrass [*Echinochloa crus-galli*]), can lead to the long-term buildup of enormous weed seed banks in the soil (Norris 1992). As a practical matter, preventing all weeds from forming seeds is not a realistic goal, but efforts to reduce seed production will reduce future weed problems.

Using an IPM Approach for Problem Weeds

Integrated pest management strategies involve understanding the life cycle of weeds: when they germinate, when they are in an active vegetative growth stage, when they set seed, and when specific control measures are most effective. Once a pest's biology is understood, you implement multiple management strategies for its control. Since weed control takes energy and time, it makes sense to try to eliminate weeds when they are most vulnerable. For example, weeds are highly susceptible to heat when they are small and tender. Flaming the area beneath the vines in December, soon after weeds emerge and reach the first true leaf stage, can have a much greater effect on reducing weed growth than if you wait until they are 6 inches tall before you try to control them. In the same way, frequent, shallow tillage beneath the vines will keep weeds from becoming very large and will also deplete the weed seed bank over time. Some weeds—nutsedge and field bindweed, for instance—do not compete well with other weeds and do not grow as well if a competitive cover crop is planted in the vineyard.

Weed Mapping

Weed distribution usually is not uniform. Patches of weeds, and particularly patches of troublesome perennials, should be managed individually. Where weeds grow in patches, you can use a global positioning system (GPS) to map the population and then treat those areas only. This sort of selective control costs less than treating the whole vineyard (Koller and Lanini 2005). If you map the weed patches, you make it easier to do follow-up evaluations and treatments. This is particularly important if a particular problem weed is just beginning to appear in the vineyard, especially if it is a species that is a candidate for eradication.

Combining Weed Control Methods

Herbicide resistance has been documented in more than 100 weed species (Warwick 1991), emphasizing the importance of developing nonchemical weed control options. While resistance to herbicides is well documented, weeds can also become resistant to other cultural practices (Liebman and Gallandt 1997). Mowing, for example, often selects for weeds that are able either to set seed prior to mowing or to set seed below the mower's cutting height. All of this means that using at least two weed control methods in combination, rather than continuously relying on a single weed control method or product, is always advantageous. The most effective combination of practices will vary from vineyard to vineyard—again depending on the weed species, the soil, and the location. Continued grower experiences and further research are needed to determine the optimal approaches for weed management in organic vineyards. For a specific weed problem, consult with your local UC Farm Advisor or a pest control adviser (PCA) to develop a strategy.

SPECIFIC WEED CONTROL TOOLS AND APPROACHES FOR ORGANIC GROWERS

Your choice of the most appropriate methods for managing weeds and other vegetation in your organic vineyard will depend on the location of the vineyard, its soil, the types of plants that you encounter, the age and health of the vines, how much money you have available, and what your weed control objectives are. The management techniques that follow are the most commonly used among organic winegrape growers.

Mulches

A mulch blocks light to prevent the germination or growth of many weed species. Many materials are used as mulches, including plastics as well as organic materials such as composted municipal yard waste, wood chips, straw, hay, sawdust, and newspaper (Porter 1999). To be effective, a mulch needs to block all light to the weeds, but the different mulch materials vary in the amount or depth of mulch that it takes to accomplish this. For effective weed control, you have to maintain a layer of organic mulch at least 4 inches thick (Makus et al. 1994).

Organic mulches break down biologically and disappear over time, typically losing about 60 percent of their thickness after one year. Usually a grower has to reapply organic mulch during the growing season to replace material that has degraded. The coarser the mulch (that is, the larger the particles in the mulch), the longer it will take to break down. However, coarser material may allow more light to penetrate and so encourage more weeds to grow, so you need to spread it carefully to get optimum shading of the soil surface (Elmore, Donaldson, and Smith 1998).

Mulches provide numerous benefits besides weed control. They conserve moisture by reducing evaporation from the soil. Soil temperature is more stable. As the mulch breaks down, it adds organic matter to the soil. Weed germination is greatly impeded and weed growth diminished by mulch. Some grasses do survive mulching, but those that do are shallow rooted and easily pulled out. Partially rotted straw or hay can be used as mulch, but because of organic certification rules it must come from fields that have not used pesticides or chemical fertilizers.

You can grow cover crops in the vine middles during the spring and then "mow-and-throw" them to create green mulch around the base of the vines. A number of cover crop species have the potential to be used to create mulches in vineyards (Bugg et al. 1996). This works well, provided that the mulch layer is sufficiently thick. Cover crops that have worked well for this technique are forage oats and cereal rye (Elmore, Donaldson, and Smith 1998). Either of these grasses can be mixed with vetch, which also creates a large quantity of biomass. Control any weeds that emerge through the mulch using an organic contact herbicide or hand weeding. Cover crops should not be planted under the vine row itself, as they may compete so much with the vines that they reduce grape yields (Colugnati et al. 2003). Unfortunately, many vineyard soils are not able to grow enough organic matter in a cover crop to make a mow-and-throw mulch a practical solution. In such cases, you can add more mulching material to augment the mow-and-throw material and make the mulch layer more effective.

Mulches may also have disadvantages in vineyards, mainly because the application of a sufficiently thick layer of mulch can be expensive and labor intensive, especially if you purchase the mulch and have it trucked in from a distance. A mulch can sometimes aggravate rodent problems, so you will want to monitor pest populations when you use mulch in a vineyard (Elmore, Donaldson, and Smith 1998). To spread mulch, you generally also have to transport very heavy loads into the vineyard, and that can contribute to soil compaction, particularly if the soil is moist at the time of application.

Some growers have had great success using woven weed-prevention fabrics, which they apply beneath vines, often at the time of planting. These materials usually come on a roll between 2 and 3 feet wide and can be mechanically applied. They are expensive, but in some cases they can provide weed control for up to 10 years. This could be a very useful option where mechanical weed control would be difficult, such as beneath vines grown on terraces.

Cultivation

Cultivation is probably the most widely used method of weed control in organic farming systems. Mechanical cultivation uproots or buries weeds. Burial works best for small weeds, while destroying the root-shoot connection by slicing, cutting, or turning the soil is more effective for large weeds. In systems that maintain permanent sod in the middles, cultivation may be limited to a strip under the drip line. The reverse might be true in systems where mulches are employed for weed suppression under the vines and cultivation is used to control weeds and incorporate cover crops between the rows. In any case, cultivation must be kept shallow to minimize damage to vine roots and to avoid bringing more weed seeds nearer the surface, where they can germinate.

Perennial weeds, because of their established root systems, are difficult to kill with a single tillage operation. In a sense, an herbaceous perennial weed is like an iceberg, where a large percentage of its mass is unseen (the underwater portion of the iceberg or the weed's root system in the soil), and only a small percentage is visible (emerged aerial portions of both). Tillage removes the top growth, and the weed regrows using its underground reserves. For perennials, a tillage 3 or 4 inches deep depletes the maximum amount of root carbohydrate reserves and forces the weeds to use up more of their reserves than with shallow tilling (Dierauer and Thomas 1994). Repeated tillage depletes the weed's carbohydrates, stored in its root system, and the plant eventually exhausts its resources and dies. This approach usually requires clean tillage of the vineyard floor. It can be an effective way to control weeds like bindweed, Bermudagrass, and nutsedge.

Although cultivation is the most common form of weed management in organic vineyards, it does have its disadvantages. Cultivation is energy intensive. Organic winegrowers sometimes make up to 16 passes in one season to keep the vineyard floor clean. Another problem is that cultivation can contribute to soil compaction as well as soil erosion, which may cause siltation in streams and rivers during wet weather. Excessive cultivation can also degrade the physical structure of some types of soil. If you use crawler tractors rather than wheel tractors and till less frequently, you can prevent this kind of soil compaction, degradation, and erosion, or at least keep it to a minimum. Limiting tillage to alternate rows is another effective way to achieve a balance between soil protection and tillage for under-the-vine weed control.

Cultivation Equipment

A wide variety of cultivation equipment is available for weed management in organic vineyards (Table 7.4). The options include many kinds of mechanical cultivators for the middles of rows and also for in-row (under-the-vine) tillage. Each grower's choice of machinery depends on the terrain (whether sloped or flat), the soil type, and the types of weeds and spacing of vines, as well as the grower's preferred approach to soil management and the availability of financial resources. Some types of equipment give the soil a deep cultivation, whereas others merely plow or hoe (or burn) the soil surface or surface layer. Mechanical hoes are also available, usually equipped with mechanical trip systems to cultivate the weeds without damaging the vines. Mechanical trip systems contact the trunk of the vine and trigger the rig to hydraulically move the cultivator away from the trunk, leaving the vine undamaged (Table 7.4).

In a vine planting where you want to conduct in-row tillage, mechanical hoes such as the Weed Badger, Kimco, or Green Hoe may be useful. These tractor-mounted cultivators extend from the tractor and can till right up to the vine without damaging the plant. Attachment options include powered rotary tillage tools and scraper blades that can move soil away from or to the base of the vines. Scraper blades have been used in vineyards for many years, where they have been called "grape hoes." Specialty tillage equipment is expensive, though, and may not be cost effective for very small vineyards.

In a vineyard less than 3 years old, the vine trunks are still slender and vulnerable to damage, but weed control is still necessary so a grow tube (a plastic tube placed around the vine trunk to protect it) and a strong stake are recommended for the young vines. Even the best cultivation equipment will not eliminate all weeds, so supplemental hand weeding is often needed. Hand cultivation by itself may be effective on a small scale.

There are also some simple articulating spring-loaded blade cultivators that can be used in fairly flat sites, as long as the soils are not too rocky. These devices simply slice the weed tops off just beneath the soil surface. They work best when weeds are small.

Row middles are usually tilled with disks or cultivators. Tandem disks are often used, as they are affordable, require relatively little maintenance, and do not displace the soil much, especially beneath the vines. Offset disks are sometimes used, but these have a tendency to build a berm on one side of the vineyard row. Disk implements take a lot of energy to operate and need a moderate amount of soil moisture to work well. If the ground is too wet, they forms clods that clog the disk blades. If the soil is too dry, the disk won't penetrate at all. Over time, disking tends to form a compacted horizon where the blades shear the soil, usually around 5 to 8 inches deep. These compacted zones can inhibit the growth of cover crops and under severe circumstances may limit water penetration. Occasional shallow ripping with short shanks can correct this problem.

Table 7.4. Machines that can be used in organic vineyards for managing vineyard floor vegetation*

Type of machine	Short description
French plow or hoe plow	Heavy duty, traditionally used, does best in moist soil. Soil must be thrown back under the vine in a subsequent pass.
Clemens cultivator	Sturdy, few moving parts, slices under the weeds, can be mounted in front, middle, or rear. Can handle tight spaces and is faster than some other cultivators.
Kimco cultivator	Extremely heavy duty, usually very slow. Can be fit with a cultivator or mower head. Some adjustment of the angle is possible, allowing use on slopes. The cultivator teeth wear out rapidly.
Gearmore cultivator	Similar to the Clemens, but not as heavy duty. It is considered reliable though lighter than the Clemens, and it uses a blade to slice weeds. Less expensive than Clemens.
Weed Badger cultivator	Fairly heavy duty, different models vary in durability. Lots of moving parts, fair reliability, but slow. Can be fit with a cultivator or mower, works for any sized weed. Head can be adjusted to work on a variety of angles and slopes but may be too wide for narrow rows.
Pellenc Sunflower cultivator	Works on single and dual rows, mid or rear mount. Good ability to handle slopes, works best on small and medium weeds. Teeth wear out rapidly, hardfacing is needed. Fairly slow and expensive.
Bezzerides cultivator	Shallow cultivator with single rod/blade, durable, better for use on mature vines. Generally used on flat land, but can be adapted for slopes. Faster than most other cultivators. Not good for young vines.
ID David cultivator	Versatile, with several attachments including mower, weed knife, cultivator, and hoe. Slow, and many parts to maintain. Good ability to adjust to different row widths, berm heights, slopes. Fairly good sensor. Will handle large weeds. Can be mounted on front or middle.
Omnas Boomerang	Cultivates with a rotary tiller head. Its mode of articulation around vines permits close cultivation. Useful for young vineyards and on small terraces. It is flexible enough to adjust to different slopes.
Spedo (for tillage or mowing)	This machine has optional attachments for under-vine tillage, a weed knife, a rotary hoe, and a mower. It offers flexibility. Distributed by Gearmore and manufactured by Spedo & Figali.
Kimco in-row mower	Mows vegetation under the vines. This is an option for vineyards that are no-till systems.
Andros Engineering in-row mower (special design)	This is a hydraulic mower that mows vegetation under the vines. Works on a spring system and is lightweight .
"Perfect" rotary mower	This rotary mower is manufactured in Holland, and it can be initially adjusted for different row sizes. It has options for mowing between the vines, and arms for mowing under the vines.
Propane flame weeders ("flamers")	Relies on propane gas burners to produce a controlled and directed flame that passes over weeds. The intense heat sears the leaves, and the weed wilts and dies 1 to 3 days later. Needs to be used on young weeds; does not work on large weeds. Flame weeders come in various models and can be fitted to tractors or used as hand-held models. Moves slowly and may be a fire risk if used in summer.

* This list may not be complete, but it includes many types of equipment used in vineyards. The inclusion of brand names and manufacturers does not indicate endorsement; the brands are intended for identification purposes only. The list is based on information from ATTRA, Organic Ag Advisors, Tom Piper and Pebble Smith, and Fetzer Vineyards, and from equipment manufacturers. This summary was prepared by Ann Thrupp, and versions have appeared previously in other publications.

A cultivator requires less horsepower than a disk and can be used to break up clods and pulverize the soil following initial disking. There are many types of cultivators, including sweeps, tines, and others. These implements also tend to conserve more moisture than disks, since they do not invert or mix the soil as much.

Rotary cultivators (Rototillers) require a large amount of horsepower to operate: 40 horsepower or more. They destroy the soil structure, so their use should be limited to preparing vineyard soils for planting where grading, incorporation of soil amendments and fertilizers, and complete weed elimination are important.

A weed knife consists of a heavy bar with a series of blades that skim the surface of the cultivated vineyard floor. It is a tool growers use when clean cultivation is desired. The cut weeds, however, continue to grow vegetatively. Weed knives are relatively inexpensive and do not require much horsepower: only 10 to 20 horsepower. They do not invert the soil, and their use helps to minimize moisture loss, especially in vineyards that are farmed without irrigation. Weed knives can be very useful for control of field bindweed: frequent, shallow tillage greatly discourages this troublesome weed.

Mowing

Mowing weeds (and cover crops) is another option for weed management in vineyards and is common in vineyards that use a non-tillage or low-tillage approach. This is sometimes a preferable method for managing the naturalized vegetation, especially in vineyards that are on hillsides or where the vines are very vigorous. Once again, a variety of equipment is available for mowing vineyard rows (see Table 7.4). Usually in these low-till mowed systems, growers use some form of weed removal under the vines (in-rows), such as cultivation, but some growers even use under-vine mowers instead of cultivation. Mowing may be less appropriate for vineyards with weak or low-vigor vines, since naturalized vegetation can create competition for moisture and nutrients and accentuate the vines' weakness.

Growers typically use two types of mowers:

- *Rotary mowers* have large blades that rotate parallel to the ground. They coarsely chop and cut prunings and plants that are growing on the vineyard floor. These mowers use less horsepower than other types and are not expensive to maintain. They can be mounted on the front or back of a tractor.

- *Flail mowers* have numerous small blades that rotate perpendicular to the soil surface. It is possible to mow to the soil level with these machines. They use more horsepower than rotary mowers, but they can grind prunings and weeds to small pieces. They have less range of adjustment for mowing height than rotary mowers. Flail mowers are usually mounted in the rear of the tractor.

Cover Crops for Weed Suppression

Cover crops can be used to suppress weed growth, since cover crops tend to reduce the overall density of weed species in the vineyard. A densely seeded cover crop reduces weeds more than a sparse stand of cover crops. Some of the most effective cover crops for weed suppression are winter cereals such as oat, barley, wheat, rye, and triticale, because they all grow faster than weeds.

Other cover crops, including subterranean clover, vetch, peas, and bur medic, grow well in cool weather and can compete with winter weeds. A mixture of cereals and legumes helps to further reduce the light available for weed growth. The optimum time to plant is in late summer or fall. Perennial cover crops such as white clover, perennial ryegrass, or orchardgrass can also help suppress weeds, but these plants are generally very competitive and may not be desirable in low-vigor vineyards. They work best in a vineyard with deep soil and can compete very effectively with nutsedges, bindweed, and many summer annual weeds (McGourty and Christensen 1998, Elmore, Donaldson, and Smith 1998, Bugg et al. 1996). Cover crops have numerous additional benefits, such as enhancing organic matter, preventing erosion, and improving water infiltration, soil fertility, and soil structure, and adding nitrogen to the system (when the cover crop includes a legume).

Organically Acceptable Herbicides

Several organic, contact-type herbicide products are available for weed control. These include essential oil products such as Matran EC, GreenMatch EX, and Weed Zap, and acetic acid/citric acid products such as AllDown and C-Cide. All of these products damage any green vegetation that they contact, including the leaves and young stems of grapevines. However, they are safe as directed sprays against older woody stems and trunks. Because these herbicides only kill the tissue they contact, thorough coverage is essential. A high application volume and addition of an organically acceptable adjuvant are recommended. It is essential that you apply these herbicides when the weeds are young or newly emerged, as the products lack effectiveness when weeds are large. These materials have no residual activity, so you will need to make repeated applications to control new flushes of weed growth.

Another organic herbicide that has appeared on the market in the last few years is corn gluten meal, which is sold under many trade names (Bingaman and Christians 1995). It is expensive and has failed to provide even minimal weed control in the vast majority of California trials.

Organic herbicides in general are expensive at this time and may not be affordable for commercial vineyards. In addition, organic herbicides are weak on grasses and on weeds with waxy or hairy leaves and stems. Finally, you should always seek approval from your organic certifier in advance, since the use of such alternative herbicides is not cleared by all agencies. The efficacy of all these materials is much lower than synthetic herbicides.

Water Management

Water management can be a key tool for controlling weeds in vineyards, especially during the summer months. Drip lines buried below the soil surface can provide moisture to the vines and minimize the amount of moisture available to weeds (Grattan, Schwankl, and Lanini 1988; Biasi et al. 2000). If properly managed, this technique can improve weed control during the dry periods of the year. Possible disadvantages include the need for (and difficulty of) quick detection of any malfunctioning or blocked emitters, the potential for damage to buried lines from roots and rodents, and the difficulty of ensuring accurate placement of the line to ensure optimum irrigation.

Overhead sprinkler irrigation promotes weed growth on the vineyard floor. To keep such vineyards weed-free, growers use cultivators or weed knives to control emerging weeds. Additional mowing may be required in vineyards under no-till management. Drip systems have largely replaced sprinklers for irrigation during the growing season. Sprinklers are used mostly for frost protection or summer heat control and to germinate cover crops and irrigate vines after harvest.

Drip-irrigated vineyards may require additional passes with under-the-vine weed cultivators or mowers compared to other irrigation methods, due to the increased frequency of irrigation. Normally, drip-irrigated vineyards are cultivated or mowed about once a month during the time that the vines are being irrigated. Left unchecked, the weeds will grow large and interfere with under-the-vine weed control the next season.

Grazing Animals

Some growers use grazing animals for vineyard weed management. In biodynamic winegrowing, integration of animal and plant agriculture is a goal, as practitioners believe that there is a complementary effect when animals are integrated directly into plant agriculture. Biodynamic growers may use chickens, rabbits, sheep, turkeys, and geese in their vineyards, usually on a small scale, to help with both weed and insect control and to diversify their farm enterprises.

Animals serve an additional purpose in vineyards: they provide manure, which can reduce the need for applications of compost or other organic nutrients. Managed intensive grazing can remove weeds from cover crops and help promote a denser cover crop stand. This kind of grazing may require additional fencing to keep the grazing animals where you want them and to facilitate rotation of their grazing. Lightweight electric fencing inside of the vineyard is easy to move and cost effective. Stouter perimeter fences may be required in some areas to prevent predation by wildlife (such as coyotes) and dogs.

Animals require special attention and facilities, including fencing, watering, roosting or nesting places, and supplemental feeding. Regular attention is required to ensure that the animals are receiving proper care. They can be prone to diseases and other conditions, and hazards that may cause them to die with little warning. More experience and research are needed to assess the costs and benefits of grazing animals in vineyards, but they appear to present a promising option for weed management in many situations.

Geese. Geese have been used for weed management in several crops for many years. All breeds of geese will graze weeds. There is literature citing their use in strawberries and occasional mentions of their use in tree orchards. One article (Clark et al. 1995) noted that geese were observed to feed heavily on weeds—especially grasses—in an apple orchard, reducing weed populations there by more than 90 percent.

Geese prefer grass species and will eat other weeds and crops only when they are hungry after the grasses are gone. If confined, they will even dig up and eat Johnsongrass and Bermudagrass rhizomes. They appear to prefer Bermudagrass and Johnsongrass, especially troublesome weeds in vineyards.

Avoid placing geese near any grass crops (i.e., corn, sorghum, small grains, etc.), as this is their preferred food. Geese also require drinking water, shade during hot weather, and protection from dogs. Portable electric fencing helps confine them to the area to be weeded and helps keep dogs and other predators out. Young geese (less than 1 year old) work best, as their major interests are eating and sleeping.

Sheep. Sheep are also used to graze weeds in vineyards. Sheep can remove all vegetation to ground level if forced to graze long enough on the same spot. If the sheep graze low-growing cover crops such as pasture species like bur medic or subterranean clover, you should manage them to intensively graze small areas for limited periods of time. When allowed to graze a large area, sheep will eat the most palatable plants and leave untouched any plants that are bitter (often the very weeds you want to control). When confined to a small area for a short time period, sheep tend to eat and trample everything quickly. Because forage clovers and grasses have buds close to the ground, they recover quickly from grazing and can then compete strongly for space with other, slower-recovering weeds.

Sheep have been used effectively for vineyard floor management mainly in the North Coast region, where many vineyards have deer fencing. The fencing keeps deer out but also protects the sheep inside from wild predators and domestic dogs. Sheep are especially useful for managing weeds in steep-sloped hillside vineyards, which can be relatively hard to cultivate or mow using heavy equipment, especially during the wet portions of the year. Sheep are typically grazed during the vineyards' dormant period, starting in December or January and continuing for a few months until the buds start to emerge on the grapevines. This corresponds with lambing season, and shepherds like the security of deer fencing and interior electric fences for the benefit of their ewes and lambs.

Most breeds of sheep need to be removed from the vineyard before budbreak because otherwise they may eat the new green buds. However, a unique breed of miniature sheep called South Down or Baby Doll Sheep can be used to graze for a longer time in the vineyards, beyond budbreak, because they are generally shorter than the trellis and are less likely to eat the buds or vine shoots. They usually prefer to eat grasses and forbs, so they focus on grazing the vineyard floor.

Goats are not recommended for weed control in vineyards since they are browsers (they prefer to eat woody stems) and are likely to chew the vines themselves. They can be very useful for brush control in areas outside of the vineyard and will browse species like poison oak and Himalayan blackberry vines. In some areas, you can rent sheep or goats for this purpose. Portable electric fencing, water, and protection from predators must also be provided if you want to use this strategy for vegetation management.

Flame Weeding

Flamers can be used for weed control, propane-fueled models being most common. Heat from the flamer causes the weeds' cell sap to expand, rupturing the cell walls. This occurs in most plant tissues at about 130°F (Ascard 1998; Ferrero et al. 1994). Weed burners are most effective on weeds with fewer than two true leaves (Ascard 1995). Grasses are harder to kill by flaming because their growing point is below the ground. After flaming, weeds that have been killed change from a glossy appearance to more of a matte finish, usually very rapidly. Treated foliage that retains a thumbprint mark after you squeeze it between a thumb and finger is considered to have been adequately flamed. Flaming can typically be done at 3 to 5 mph, depending on the heat output of the unit. Repeated flaming can also suppress perennial weeds, such as field bindweed. Care must be taken not to ignite dry vegetation, as there is the potential to injure vines or start a wildfire. Flaming can also damage irrigation tubing on the soil surface.

The angle, pattern, and length of the flame vary with equipment and manufacturer's recommendations, but they range from 30° to 40° below horizontal, 8 to 12 inches above the base of the weed plants, with flame lengths of approximately 12 to 15 inches. Best results are obtained when there is no wind, as winds can prevent the heat from reaching the target. Early morning and evening are the best times to see the flames so you can make adjustments. Flame Engineering, Inc. and Thermal Weed Control Systems, Inc. manufacture both hand-held and tractor-mounted flame-weeding equipment (see Table 7.4).

Drawbacks to flame weeding include its high rate of energy consumption (and associated cost), the low driving speed that makes it a time-consuming process, and irregular patterns of weed control. Additionally, many air pollution control districts discourage open-air burning of any sort or require a permit for use of a flamer. It is not unusual for dead leaves, weeds, and other debris on the vineyard floor to burn or smolder during flaming. Consequently, in many areas of California flame weeding can only be done on approved burn days, if at all.

In a study evaluating weed control by flaming (Ascard 1995), the most important variables were species and growth stage. *Malva neglecta* (cheeseweed) was resistant to flaming, which yielded little or no control of the weed. A single-pass flame treatment was insufficient to control annual weeds at later developmental stages (> 6 true leaves). For *Chenopodium album* (lambsquarters), three treatments at weekly intervals were necessary to achieve 95 percent control. The weed's developmental stage was also crucial to the effectiveness of *Taraxacum officinalis* (dandelion) control by flaming. Small plants were killed by a single flaming, whereas bigger plants often survived four flamings (only 69% control). Where flame-tolerant or perennial weeds occurred in the vineyards, four treatments only achieved a weed reduction of 76 percent. In a vineyard with mostly annual weeds, three treatments controlled 95 percent of the weeds.

Hand Weeding

Hand weeding is practiced by nearly all organic winegrowers to clean up after they use other weeding techniques. Most mechanical weeding implements do not completely control all of the weeds around posts and stakes. Hand weeding ensures complete control and helps prevent additions to the soil's weed seed bank.

There are many different tools, such as hoes and shovels, for hand weeding. Gasoline-powered string trimmers are also useful for weeding areas that are difficult to reach with tillage equipment.

Hand weeding gives excellent results, but it can be quite expensive. As with most weeding techniques employed by organic winegrowers, hand-weeding when the weeds are relatively small works best, even if you have to make several subsequent passes. Many growers limit their hand weeding to areas where aesthetics are important, such as around tasting rooms, wineries, or places where visitors are received.

CONCLUSION

As an organic winegrower, you need to develop weed control programs that are specific for your individual vineyard, based on the type of weed species present, your own aesthetic standards, the amount of tillage you consider to be acceptable, the equipment that you own, and your available financial resources. Most organic winegrowers have developed effective and affordable programs that do not negatively affect yield or winegrape quality. Experience has shown that weed control is both time consuming and expensive for most organic producers, but that there are many tools and weed- control strategies available to successfully manage weeds in vineyards.

REFERENCES/RESOURCES

Ascard, J. 1995. Effects of flame weeding on weed species at different developmental stages. Weed Research 35(5):297–411.

———. 1998. Flame weeding: Effects of burner angle on weed control and temperature patterns. Acta Agriculturae Scandinavica 48(4):248–254.

Biasi, W., C. Peratoner, P. Gasparinetti, T. Maschio, and G. Teot. 2000. Underground drip lines for modern vineyard management. Informatore-Agrario 56(47):85–94.

Bingaman, B. R., and N. E. Christians. 1995. Greenhouse screening of corn gluten meal as a natural control product for broadleaf and grass weeds. HortScience 30:1256–1259.

Bugg, R. L., G. McGourty, M. Sarrantonio, W. T. Lanini, and R. Bartolucci. 1996. Comparison of 32 cover crops in an organic vineyard on the north coast of California. Biology, Agriculture, and Horticulture 13:63–81.

Clark, M. S., S. H. Gage, L. B. DeLind, and M. Lennington. 1995. The compatibility of domestic birds with a nonchemical agroecosystem. American Journal of Alternative Agriculture 10(3):114–120.

Colugnati, G., G. Crespan, D. Picco, F. Bregant, I. Tonetti, A. Gallas, and A. Altissimo. 2003. Performance of various species for vegetation cover in the vineyard. Informatore-Agrario 59(13):55–59.

Dierauer, H. U., and J. M. Thomas. 1994. Efficacy of different nonchemical methods of controlling broadleaf dock (Rumex obtusifolius). Maîtrise des adventices par voie non chimique. IFOAM, Dijon, France, 5–9 July 1993. 2:311–314.

Elmore, C., D. Donaldson, and R. Smith. 1998. Weed management. In Cover cropping in vineyards, C. Ingels et al., eds. Oakland: University of California, Division of Agriculture and Natural Resources, Publication 3338.

Ferrero, A., P. Balsari, G. Airoldi, and J. M. Thomas. 1994. Preliminary results of flame weeding in orchards. Maîtrise des adventices par voie non chimique. IFOAM, Dijon, France, 5–9 July 1993. 2:389–394.

Grattan, S. R., L. J. Schwankl, and W. T. Lanini. 1988. Weed control by subsurface drip irrigation. California Agriculture 42:22–24.

Johnson, C. 1960. Management of weeder geese in commercial fields. California Agriculture, August 1990, p. 5.

Kolberg, R. L., and L. J. Wiles. 2002. Effect of steam application on cropland weeds. Weed Technology 16(1):43–49.

Koller, M., and W. T. Lanini. 2005. Site specific herbicide application based on weed maps. California Agriculture 59(3):182–187.

Liebman, M., and E. R. Gallandt. 1997. Ecological management of crop-weed interactions. In Ecology in agriculture, L. E. Jackson, ed. San Diego: Academic Press. Pp. 291–343.

Lisa, L., S. Parena, L. Lisa, and C. Lozzia. 2000. Comparison of different techniques of soil management and weed control in the northern Monferrato area. Proceedings of the Meeting on Integrated Control in Viticulture. 23(4):209–211.

Makus, D. J., S. C. Tiwari, H. A. Pearson, J. D. Haywood, and A. E. Tiarks. 1994. Okra production with pine straw mulch. Agroforestry Systems 27:121–127.

McGourty, G., and L. P. Christensen. 1998. Cover cropping systems and their management. In Cover cropping in vineyards, C. Ingels et al., eds. Oakland: University of California, Division of Agriculture and Natural Resources. Publication 3338. Pp. 43–57.

Norris, R. F. 1981. Zero tolerance for weeds? Proceedings of the California Weed Conference 33:46–49.

———. 1992. Case history for weed competition/population ecology: Barnyardgrass (Echinochloa grus-galli) in sugarbeets (Beta vulgaris). Weed Technology 6:20–227.

Porter, C. 1999. California wineries take major steps to improve vineyards. BioCycle 40(1):59–62.

Scopel, A. L., C. L. Ballare, and S. R. Radosevich. 1994. Photostimulation of seed germination during soil tillage. New Phytologist 126(1):145–152.

Tworkoski, T. 2002. Herbicide effects of essential oils. Weed Science 50(4):425–431.

Warwick, S. I. 1991. Herbicide resistance in weedy plants: Physiology and population biology. Annual Review of Ecological Systems 22:95–114.

Zabadal, T. J., and T. W. Dittmer. 2001. Influence of weed control, nitrogen fertilization, irrigation, and pruning severity on the establishment of Niagara grapevines. Small Fruits Review 1(31):21–28.

8 Organic Grapevine Disease Management in California

W. DOUGLAS GUBLER AND JANET C. "JENNY" BROOME

Plant disease management for organic wine grapes is a continuing challenge. However, with careful forethought and planning in vineyard establishment, including appropriate choice of cultivars and rootstocks, row orientation, and trellis systems, and diligent use of clean, certified planting stock (rootstock and scion), organic growers should be able to avoid certain problems. Proper attention to good viticultural practices that result in adequate vine vigor for the site, together with favorable environmental conditions and timely, appropriate canopy management practices, can improve wine quality while reducing disease problems. Nevertheless, some diseases such as powdery mildew are universally present in California vineyards and require a systematic, preventative approach to treatment every year. Organic disease management practices should be undertaken and integrated with other cultural and pest management practices and key actions that are taken at appropriate times throughout the growing and dormant seasons. In this chapter, we describe the most important diseases affecting wine grape production, including symptoms, pathogen life cycles, and management options. At the end of the chapter, we present a calendar of activities (see Table 8.4) that lays out the annual growing cycle and key times for disease management interventions, including related relevant cultural practices.

In order to produce certified organic wine or wine made with organically grown grapes, one must be certified by an accredited organization. The certification process involves developing an organic system plan, complying with all rules regarding management practices and inputs, and keeping proper records. Any disease management materials used in organic production must be allowed by your certifier. The general rule is that the USDA's National Organic Program (NOP) allows natural (nonsynthetic) substances and prohibits synthetic substances unless they appear on the National List. The National List contains the names of allowed synthetic and prohibited natural substances that are exceptions to the general rule. A material that is listed for a specific use is also restricted to that use (i.e., a soap that is listed for insect control is not allowed for disease control). Some materials have specific restrictions; for instance, copper must be used in a manner that minimizes its accumulation in the soil. Growers are required to document all products applied to crop and soil. You have to know all of the ingredients in a product in order to determine whether or not it complies with the NOP. Certifiers either review products themselves or rely on an outside service.

FOLIAR AND FRUIT FUNGAL PATHOGENS AND DISEASES

Powdery mildew (*Erysiphe necator*)

Powdery mildew (*Erysiphe necator,* syn. *Uncinula necator*) is the most significant disease affecting grapes in California (Gubler and Hirschfeldt 1992). The pathogen, an obligate parasite, is native to North America and thus co-evolved with native *Vitis labrusca* grape cultivars. However, when the much more susceptible *Vitis vinifera* European winegrape cultivars were introduced to California, powdery mildew became a severe problem. The fungus penetrates the cells of the epidermis of leaves, green shoots, and juvenile berry tissue and may reduce photosynthetic rates. Severe damage occurs because infected berries are stunted and cracked, leading to reduced sugars and thus reduced quality and increased bunch rot. Powdery mildew probably is present to some degree in virtually every vineyard every year. Severity of infection varies among vineyards and is dependent on the grape cultivar, how early in the season disease onset occurs, and weather conditions during the season. Approximately 95 percent of grape acreage is treated with fungicides for powdery mildew each year.

Symptoms. Powdery mildew is very difficult to detect in the early stages of development. On dormant canes, reddish brown blotchy areas show where infections occurred during the previous growing season. In spring, infection is followed by chlorotic spots that appear on the upper leaf surfaces and later by white, weblike mycelium that grows superficially on the lower leaf surface, below the chlorositic spots. Still later, the fungus produces powdery white spores and infection can spread to the leaves' upper surface. Berries and rachises can also become infected. If the disease on berries is not controlled, they will not develop fully and may crack. Fungal growth (mycelium) is white and weblike, and the mycelium produces conidiophores, which then produce the asexual spores, or conidia, continuing the epidemic by infecting current-season host tissue. This white webby growth on leaves and fruit is visible to the naked eye.

Pathogen biology and disease cycle. In most of California, the fungus overwinters principally as ascospores in small black fruiting bodies (chasmothecia) (Gubler and Hirschfeldt 1992) (Figure 8.1). Chasmothecia reside on the bark, canes, and spurs where they have washed from leaves during fall and winter rains. Ascospores produced in these fruiting bodies require free moisture for release, germination, and infection. They are released in the spring and forcibly ejected from the chasmothecia. They alight on the lower surfaces of basal leaves during rainfall, sprinkler irrigation, or heavy fogs and dews. High relative humidity (RH) conditions increase sporulation and lesion expansion on both fruit and foliage. The optimum RH for this pathogen is 65 percent, but this is primarily important only during periods of optimum temperature (70° to 86°F) (Rea and Gubler 2002). Ascospores are released when temperatures are at 50° to 86°F, with optimum releases and greatest efficiency of germination and infection at 68° to 75°F. Germination and infection are reduced when temperatures go above 77° or below 68°F.

The other overwintering form for grapevine powdery mildew in California is bud perennation (Sall and Wyrinski 1982; Ypema and Gubler 2000).

When the buds with this condition push in the spring, infected tissue emerges. Bud perennation occurs statewide but is most common on the Carignane variety in the Central Valley. The condition has also been observed on Chardonnay in the coastal counties and on Thompson Seedless in the Central Valley, but in these instances bud perennation followed years when the disease occurred early and was severe for the entire season. At one time, this form of overwintering was thought to be the result of a piece of the pathogen's mycelium having been trapped inside the bud at some point during the season.

Work by Rumbolz and Gubler (2005) showed, however, that the mycelium in the bud resulted in infected leaf prophyls (bud scales) and trichomes (leaf hairs) within the bud. Fungal infection structures called haustoria were evident in many of these cells. These haustoria become the overwintering structure, even though they are attached to the mycelium. Rumbolz and Gubler also found both germinating and nongerminated conidia within the dormant bud. When buds push in the spring, the

Figure 8.1. Primary inoculum of powdery mildew *(Erysiphe necator)* in grapes: (A) cleistothecium releasing ascospores; (B and C) infected shoots from bud penetration. *Photos* by W. D. Gubler, J. K. Clark, and J. C. Broome.

infected tissue emerges from the bud and the fungus will begin to grow and sporulate immediately if the temperature is between 68° and 80°F. Temperatures above 92°F somewhat inhibit the growth of the fungus, with leaves and shoots showing no symptoms. Rumbolz and Gubler also showed that the most efficient time for bud infection to occur was the 3 to 6 true leaf stage. When disease onset occurred during that period, the infection rate in buds was significantly higher than when infection occurred at budbreak or at 9 true leaves. This work also showed the stage at which buds became infected in a given year to be the same as the stage at which disease onset occurred the following season. By knowing when disease onset occurs in one year, then, a grower would know when to expect first disease to be expressed in the following year, and so would be better prepared to time fungicide applications for best effect.

Once disease onset has occurred, the epidemic is driven primarily by temperature. When temperatures are between 70° and 86°F, the pathogen can reproduce as rapidly as every 5 days (Delp 1954). Between 86° and 92°F, the fungus grows much slower, with approximately 15 days to reach reproduction (Ypema and Gubler 2000). As the fungus is exposed to higher and higher temperatures, rates of spore production, germination, and infection go down. When temperatures reach 94°F or hotter for extended periods of time, E. necator cannot grow or reproduce, so the disease is less of a problem. Free moisture has a negative effect on conidial germination and growth (Delp 1954; Chellemi and Marois 1991a). However, recent research suggests that the spores of this pathogen are capable of surviving for several hours after imbibing water before they burst (Gubler, Unpublished). Direct sunlight (in particular, ultraviolet light) can harm the fungus by killing it or slowing the spore germination and infection processes. Conversely, the fungus tends to grow unimpeded on green tissue in shaded canopy areas and on fruit that is protected from direct exposure to sunlight.

Mildew management. Early season monitoring for disease is necessary for good control. You can monitor by randomly selecting and assessing 10 to 15 basal leaves from grapevines 7 to 10 days after the first spring rainfall after budbreak. Colonies derived from ascospore infections can be found on the underside of basal leaves and appear as small

(0.5 to 1.0 cm) areas of gray-white growth. If you hold the leaves up to the sunlight, the infected areas will appear as a light yellowing of the tissue.

Powdery mildew thrives in dense canopies (Figure 8.2). Cultural controls such as trellising, cane cutting, and training techniques, as well as leaf removal at berry set, are important in limiting powdery mildew as they create a more open canopy, which improves coverage of fungicides and makes the canopy microclimate less conducive to disease (Chellemi and Marois 1992). Exposure of grape clusters to sunlight at cluster set allows the berries to develop a thicker cuticle and also

Figure 8.2. Powdery mildew thriving on fruit under the dense canopy of a vigorously growing vine. *Photo* by W. D. Gubler.

reduces the ability of E. necator to grow and develop on exposed berries. Leaf removal is usually done immediately after fruit set in the spring.

Varietal selection is also an important consideration, especially where powdery mildew is known to be a significant problem, such as in parts of the Central Coast. All V. vinifera grapes are highly susceptible. Carignane is one of the most susceptible varieties, followed by Chardonnay, Chenin Blanc, and Cabernet Sauvignon (Table 8.1). Petite Sirah, Zinfandel, Semillon, and White Riesling are somewhat less susceptible, though in years when powdery mildew is severe, even they may show considerable disease.

The use of well-managed cover crops can help you manage powdery mildew by improving your access to the vineyard under wet soil conditions for early season mildew treatment. Cover crops enable better traction in the vineyard during rainy periods early in the growing season. However, if the cover crop growth is lush or too tall, it will contribute to

Table 8.1. Varietal susceptibility *(Vitis vinifera)* to major diseases in California

Variety	Powdery mildew	Botrytis	Eutypa
Barbera	H	M	H
Cabernet Franc	H	L	H
Cabernet Sauvignon	H	L	H
Chardonnay	H	H	H
Chenin Blanc	H	H	H
Grey Riesling	M	H	H
Merlot	H	M	H
Petite Sirah	M	L	H
Pinot Blanc	H	M	H
Pinot Noir	H	H	H
Sauvignon Blanc	H	H	H
Semillon	M	M	M
Syrah	M	L	H
White Riesling	H	H	M
Zinfandel	M	H	M

H = High, M = Moderate, L = Low, ? = unknown

Sources: Gadoury 1995; UC IPM Pest Management Guidelines 2002, www.ipm.ucdavis.edu/
PMG/selectnewpest.grapes.html; Purcell 2005, *Xylella fastidiosa* web page, www.cnr.
berkeley.edu/xylella/fallsympt.html.

increased relative humidity in the vineyard canopy. When temperatures are in the optimum range for infection, higher humidity can increase both the incidence and the severity of powdery mildew. Keep the cover crop vegetation mowed short to improve airflow and reduce humidity in the vine canopy when the disease pressure of powdery mildew is otherwise high.

Organic fungicides (Table 8.2) are necessary for powdery mildew control in most vineyards in California. Preventive chemical treatments for powdery mildew are necessary from shortly after budbreak through veraison, with the exact timing of treatment dependent on the region, trellis type, cultivar, product used, and weather. By far the most popular material for powdery mildew control is sulfur. Many growers (both conventional and organic) rely exclusively on sulfur for powdery mildew control. Sulfur comes in dust and dry flowable sprayable formulations and is typically applied at intervals of 7 to 10 days for dusts and up to 18 days for dry flowable sulfur sprays. Sulfur dust applications have at least one disadvantage: they have been shown to exacerbate spider mite outbreaks by suppressing beneficial arthropods (predatory thrips and mites) in vineyards (Hanna et al. 1997). Sulfur residues

in excess of 0.4 micrograms per gram of berry weight also will cause off-flavors in wines (Thomas et al. 1993). Also, sulfur dust has recently been associated with major drift incidents in several winegrape-growing regions, so the California Department of Pesticide Regulation is looking at potential restrictions on its use. Last of all, sulfur dust can cause foliar burning when temperatures rise above 95°F and so should not be used when the weather is likely to be hot.

Sulfur products for spray applications include wettable formulations (an older type of formulation no longer recommended for use, due to its multi-particulate size and application problems) and dry flowable (df) micronized formulations. Although df sulfur is slightly more expensive than sulfur dust and takes more time to apply, it does not bring with it the problems with drift, worker health and safety, and mite outbreaks common to other formulations. The first application of micronized df sulfur should be at full budbreak, delivered in a great enough volume of water to achieve thorough wetting of the cordons and spurs or canes. This application method has been shown to consistently reduce disease at its onset by approximately 95 percent. The treatment works by artificially stimulating the release of ascospores. As there is very little foliage at this time, the spores will generally land on the ground. Those spores that do land on the leaves are killed by the sulfur. The treatment is still effective even if the actual event that triggers ascospore release (rainfall, sprinkler irrigation, or fog) does not occur until several weeks after application.

Some light summer oils have been shown to provide excellent control of powdery mildew and mites, and these can be used as a replacement treatment for micronized sulfur at budbreak. Growers should choose oils low in unsulfonated residues to keep the chances of phytotoxicity to a minimum. Two oils have been registered for organic use, and both provide excellent disease control under any type of pressure. There are also several new materials, including potassium bicarbonates and materials that induce systemic acquired resistance, that can be used as part of a mildew control program, though these products usually do not offer a very high level of control under heavy disease pressure. Additionally, there are new biological controls such as *Bacillus subtilis* and *B. pumilus*, both of which provide reasonable control under low to moderate disease pressure. Fixed coppers and foliar nutrient products have provided good disease control under light to medium disease pressure and are registered for organic production. Any of these materials has the potential to fit into a powdery mildew control program as part of a rotation scheme for organic growers, but coverage is the key issue. (See Table 8.2 for materials and Table 8.4 for application timings.)

If substantial mildew develops in a vineyard, oil application is the recommended treatment for eradication. Sulfur dust will not completely kill the pathogen once the disease is under way. Growers can also use lime sulfur as a dormant-season treatment to clean up after a particularly bad mildew year in regions where the most important overwintering form of the fungus is chasmothecia, and they can also apply it during the season at a low rate. Dormant applications of lime sulfur at 15 gallons per acre will also reduce overwintering *Botrytis cinerea* and *Phomopsis viticola* propagules, decrease *Cladosporium cladosporioides* (the cause of green berry or brown spot), and kill *Phaeoacremonium* [spp.] (*Togninia* [spp.]) and *Phaeomoniella chlamydospora*, the respective causes of Esca (black measles) and Petri disease (young esca) (Gubler et al., unpublished).

Powdery mildew infection is strongly related to temperature, and various researchers have developed temperature-driven models (Sall 1980; Chellemi and Marois 1991b; Gubler et al. 1999) that describe this relationship in mathematical terms. However, only the Sall and Gubler-Thomas models have been validated and used in California, and only the Gubler-Thomas model is currently used extensively in California and other production areas around the world. This predictive and risk assessment tool enables growers to know what the mildew population is doing in terms of population increase or decrease at any given time. The model, which uses grapevine canopy weather data, allows more precise

Table 8.2. Organically approved fungicides for powdery mildew control*

Active ingredient	Rate (per acre)	Comments
Sulfur dust	10–20 lb	Apply below 85°F
Micronized sulfur	2–5 lb	Apply below 90°F
Liquid lime sulfur	10 gal	Dormant treatment
Narrow range oils	1–2%	Excellent eradicant. Problems with sulfur/oil. 14 days between sulfur and oil.
Bacillus subtilis	4–8 lb	Biological control
Bacillus pumilus	2–4 qt	Biological control
Neem	1%	Natural plant product
Copper materials	1–2 lb	Fixed copper
Sodium bicarbonate	2.5–5 lb	Inorganic salt
Cinnamaldehyde	see label	Natural plant product
Potassium salts of fatty acids	1.5–2%	Also controls soft-bodied insects
Harpin protein	4.5–8 oz	Systemic acquired resistance
ABA	2–4 oz	Systemic acquired resistance
Chitanase	see label	Systemic acquired resistance
Monopotassium phosphate (MKP)	4–8 lb	Systemic acquired resistance/Foliar-applied nutrient
Not yet registered for disease control:		
Calcium metalosate		Foliar-applied nutrient

*Please see the UC IPM grape pest management guidelines (www.ipm.ucdavis.edu) for continually updated information on fungicides, including organic materials.

Figure 8.3. Monitoring weather data in the vineyard and using disease risk models like the powdery mildew Risk Assessment Index (RAI): (A) remote desktop access to weather sensor data via the Internet; (B) automated weather sensors collecting data in the vineyard. *Photos* by J. C. Broome.

scheduling of chemical applications (Figure 8.3 and 8.4) (Gubler et al. 1999).

The Gubler-Thomas model, also known as the UC Davis Risk Assessment Index (RAI), is outlined in Table 8.3 and explained in greater detail in the *UC Grape Pest Management Guidelines* (UC Statewide IPM Program 2006). It is used to help determine when infection risk is high (index of 60 to 100), a condition that prompts the grower to shorten the interval between treatments and to switch from less effective to more effective materials. If the risk is moderate (index of 40 to 50), the model prompts growers to lengthen the spray application intervals. When the index is low (index of 0 to 30), applications can be terminated. Risk is higher when temperatures are between 70° and 86°F for at least 6 hours per day and the index is 60 to 100, indicating that the pathogen is reproducing every 5 days. An index of 40 to 50 indicates a reproductive rate of every 15 days, and an index of less than 30 indicates that the pathogen is incapable of reproducing or

Figure 8.4. Temperature-driven UC Davis grape powdery mildew Risk Assessment Index (RAI), extrapolated using GIS throughout the Napa Valley.

causing disease. Organic products should be used on 5- to 7-day treatment intervals under high disease pressure, except for oils, which can be used at 1 to 2 percent volume oil/volume water on 14- to 17-day intervals (Dell et al. 1998). Sulfur dust use was successfully stretched to 30-day intervals using the Gubler-Thomas model in recent research on the North Coast (Gubler et al., unpublished).

Botrytis bunch rot *(Botrytis cinerea)*

Botrytis bunch rot, caused by the fungus *Botrytis cinerea*, can infect grape leaves, shoots, flowers, and berries (Marois, Bledsoe, and Bettiga 1992). It can be a serious problem in many premium winegrape production areas. Historically, conventional treatment for this disease consisted of the application of fungicides, often with little success, and particularly so in years when the disease was unusually severe. Since 1987, however, cultural controls have proven quite effective and have become the primary method of Botrytis control for most growers of winegrapes and table grapes.

Symptoms. Early season shoot, leaf, and flower blight may occur following spring rains. Brown lesions appear on affected tissue, followed by sporulation on the lesions, which then appear fuzzy and gray-brown (Figure 8.5). Infected fruit exhibit a browning of the epidermis, which softens quickly. The infected area of the epidermis slips off at the slightest touch, and the flesh under the lesion is rotted so that a finger will penetrate easily. This symptom has given the infection the name "slip skin." Under high humidity, this tissue will also support extensive sporulation in the form of gray-brown, velvety masses.

Pathogen biology and disease cycle. Botrytis cinerea is considered to be a weak pathogen but a very good colonizer of dead and senescent tissue. It will survive on many hosts either dead or alive. In vineyards, the fungus overwinters in mummified berries and on dormant canes as structures called sclerotia

Table 8.3. The powdery mildew Risk Assessment Index (RAI)–Gubler-Thomas model*

Starting the mildew model:

Once mildew is observed, an epidemic is predicted to begin when there are 3 consecutive days with 6 or more continuous hours of temperatures between 70° and 85°F, measured from within the vine canopy.

1. Starting with the index at 0 on the first day, add 20 points for each day with 6 or more continuous hours of temperatures between 70° and 85°F.

2. Until the index reaches 60, if a day has fewer than 6 continuous hours of temperatures between 70° and 85°F, reset the index to 0 and continue.

3. If the index reaches 60, an epidemic is under way. Begin using the spray-timing phase of the index.

Subsequent spray timing:

Each day, starting on the day after the index reached 60 points during the start phase, evaluate the temperatures and adjust the previous day's index according to the rules below. Keep a running tabulation throughout the season. In assigning points, note the following:

- If the index is already at 100, you don't add points.

- If the index is already at 0, you don't subtract points.

- You can't add more than 20 points a day.

- You can't subtract more than 10 points a day.

1. If fewer than 6 continuous hours of temperatures between 70° and 85°F occurred, subtract 10 points.

2. If 6 or more continuous hours of temperatures between 70° and 85°F occurred, add 20 points.

3. If temperatures reached 95°F for more than 15 minutes, subtract 10 points.

4. If there are 6 or more continuous hours with temperatures between 70° and 85°F AND the temperature rises to or above 95°F for at least 15 minutes, add 10 points. (This is the equivalent of combining points 2 and 3 above.)

*This model works best if you purchase a weather monitoring device that has the model built into it. Most of the major companies have incorporated the model into their machines, and it is a simple matter to set one of these up in the vineyard. We do NOT recommend that you try to use the model by keeping up with the temperature by hand. When this has been tried, even by research scientists, they have failed somewhere during the season. There is equipment available for $200 to $375 that will do a remarkably good job of calculating the model.

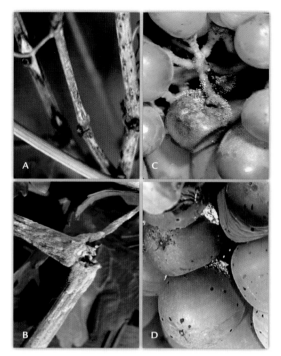

Figure 8.5. Steps in the disease cycle for Botrytis bunch rot *(Botrytis cinerea)* of grapes: (A) mycelium and sclerotia overwinter on canes and in old fruit clusters; (B) spores infect leaves and shoots in the spring; (C) spores inside a grape cluster; (D) color change of infected berries. *Photos by J. C. Broome, W. D. Gubler, and J. K. Clark.*

(Figure 8.5). With spring rains, sclerotia germinate and produce gray-brown spores (conidia). During bloom, floral tissue becomes infected during periods of moisture from rainfall, overhead sprinklers, or prolonged dew periods. This infection carries over as late-season infected blossom debris, which can become trapped in the cluster. When subsequent periods of wetness occur, this debris will contribute inoculum that will lead to berry rot.

Latent or quiescent infections can also occur when flower tissue becomes infected during bloom. The fungus remains within the berry until the sugar concentration increases, and then resumes growth and spreads throughout the berry, causing leakage of berry content. The fungus then will grow onto the sugary substrate and cause other berries to rot. Mature berries can become infected either directly, when fungal spores land on the berries, or via nesting, when the fungus grows from one infected berry into its neighbor in the cluster. Botrytis spore germination and infection require free moisture from rain or dew and proceed most rapidly within an optimal temperature range of 65° to 72°F (Figure

8.6). It can still grow within a wider temperature range of 32° to 86°F, but it will grow more slowly. High relative humidity (greater than 90%) is required for spore production. While spores can directly infect intact berries, they also enter through wounds caused by insects, birds, mechanical damage, or damage from powdery mildew. The risk of berry infection increases with increasing berry sugar.

Bunch rot management. Good cultural control of Botrytis bunch rot has been achieved using canopy management—in particular, leaf removal (Figure 8.7) of the four or five basal leaves (the leaves around the clusters) at berry set or when berries are no larger than "BB-size" (Gubler et al. 1987; Stapleton et al. 1990; Stapleton and Grant 1992). Leaf removal has also been shown to reduce sour rot, leafhopper populations, and powdery mildew (Stapleton et. al 1990; Chellemi and Marois 1992). Although this practice is labor intensive and can be expensive, it has been shown over the years to be the best treatment for this disease and to result in increased quality factors in fruit, including higher malic acid and titratable acidity, decreased potassium, increased Brix, increased color, and better wine quality (Bledsoe, Kliewer, and Marois 1988). Leaf removal has been successfully used with Chardonnay, Zinfandel, Chenin Blanc, Johannesburg Riesling, and other varieties. Other cultural techniques, such as shoot positioning, hedging, and the use of trellis types that open up the canopy, offer some degree of protection against *B. cinerea* as well. For example, in an area where vine vigor might be

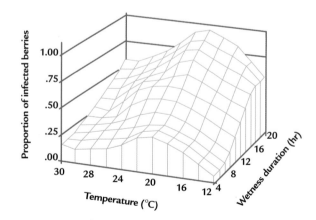

Figure 8.6. Surface response graph showing the interaction of hours of free water and temperature on berry infection by *Botrytis cinerea*. From Broome et al. 1995.

Figure 8.7. Leaf removal being performed on vines supported by a two-wire vertical shoot-positioning trellis (VSP). *Photo* by W. D. Gubler.

anticipated to be moderate to low, the combination of particular trellises (say, a two-wire vertical trellis with shoot positioning wires) and appropriately timed leaf removal will provide a canopy microclimate less conducive to disease. In addition, in new plantings, rows should be oriented to meet and take best advantage of prevailing winds.

Leaf removal and other cultural management practices around the clusters can physically shake off some of the infected floral debris. Removal of blossom debris by blowing air into the vine was documented to reduce Botrytis bunch rot by nearly 30 percent in Chardonnay and Chenin Blanc (Gubler, unpublished). Some growers run an air-blast sprayer through the vineyard to blow off debris before berry closure.

In general, proper control of vine vigor through the use of a rootstock that is appropriate for the soil type, together with adequate but not excessive irrigation and nutrients (nitrogen in particular), can reduce bunch rot problems by controlling canopy growth and thus avoiding creation of a disease-conducive canopy microclimate. These cultural practices also contribute to a more disease-tolerant grape berry due to microclimate conditions (more sunlight, more air movement) and the effects on the berry's epicuticular wax layer.

Some varieties are more susceptible than others: Chardonnay, Chenin Blanc, Pinot Noir, Sauvignon Blanc, and Zinfandel are highly susceptible (Table 8.1), whereas Cabernet Sauvignon and Cabernet Franc are less so. Tight-clustered varieties and clones tend to have more bunch rot (Vail et al. 1998; Vail

and Marois 1991). Vail found Cabernet Sauvignon berries to be quite susceptible, but one rarely sees the disease on that variety in the vineyard because its cluster architecture is so loose. It is reasonable to conclude that any practice that results in looser clusters, such as proper irrigation or use of plant growth regulators or the selection of looser-clustered clones, can reduce bunch rot. Some researchers and growers have found that an application of a hormone, gibberellic acid (GA), at bloom can extend cluster length and thereby reduce cluster tightness and Botrytis bunch rot, particularly on Zinfandel in the San Joaquin Delta region. Because of the danger that this growth hormone will reduce yields, its use is restricted to certain regions and only with a special local permit.

Chemical control of bunch rot can be important in coastal regions with wetter climates. There are four key treatment periods if wet weather conditions occur: bloom, pre-close, veraison, and pre-harvest. Preventive treatments are commonly applied at bloom, pre-close, and veraison, but only if wet conditions occur. Monitor for Botrytis by looking for grey mold symptoms on leaves, shoots, flowers, and clusters. Most fungicide treatments are based on longer-term prevention strategies, but you can also monitor weather conditions to estimate the immediate risk of infection and time chemical treatments based on the temperature and moisture requirements of the fungus (Figure 8.6; Broome et al. 1995). Unfortunately, the only organic fungicides currently available are contact materials, and they do not provide any real "kickback" activity—that is, they cannot eradicate the fungus after infection has begun. For more details on the model, see www.ipm.ucdavis.edu/DISEASE/DATABASE/grapebotrytis.html.

There are very few organic fungicides available for Botrytis, but one alternative is JMS Stylet Oil, which has been demonstrated to control disease as well as the commonly used conventional fungicides (Dell et al. 1998). Dust formulations of copper products have controlled disease in susceptible varieties such as Riesling and Chardonnay, and they also inhibit sporulation. There is also a newly registered microbial pesticide, *Bacillus pumilis*, which has some action against Botrytis.

Botrytis bunch rot, like all plant diseases, will only occur if a susceptible host plant and pathogen are present in a conducive environment, a concept known as the plant disease triangle (Figure 8.8).

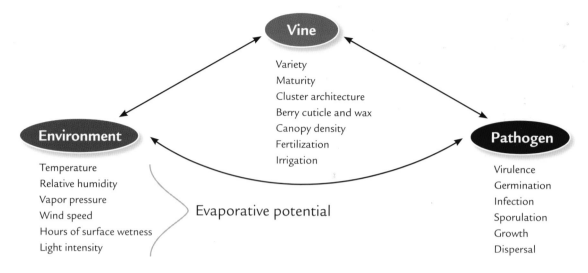

Figure 8.8. Plant disease triangle for Botrytis bunch rot of grapes.

Various management practices can influence this triangle, leading to more or less incidence and severity of disease in a particular vineyard. Organic vineyard managers must keep these balances in mind: canopy management practices can affect the microclimate, environmental conditions, and the host, and have the potential to reduce disease if properly applied.

Summer Bunch Rot Complex (*Sour Rot*)

A summer bunch rot complex (sour rot) may occur when secondary invaders take advantage of mechanical damage to berries. Fungal organisms involved include *Aspergillus niger, A. carbonarius, Alternaria tenuis, Botrytis cinerea, Cladosporium herbarum, Rhizopus arrhizus, Penicillium* spp., and others (Marois, Bledsoe, and Bettiga 1992; Rooney-Latham et al. 2009). This disease is more likely to occur in warm growing regions such as the San Joaquin Valley, but its *Aspergillus* component can, and often does, result in disease on Zinfandel in the North Coast and Delta production areas.

Symptoms, pathogen biology, and disease cycle. Masses of blue, black, or gray-brown spores develop on the surface of infected berries. Summer bunch rot often culminates in sour rot, especially in the southern San Joaquin Valley (Figure 8.9). Apart from the fungi, sour rot is caused by a variety of microorganisms, including *Acetobacter* bacteria, which are spread by vinegar flies attracted to the rotting clusters. Sour rot produces a characteristic vinegary

smell within the canopy. As veraison occurs, injured fruit becomes increasingly susceptible to invasion by a wide variety of naturally occurring microorganisms. In table grapes, two of the first fungi to invade the injured tissue are *Aspergillus niger* and *A. carbonarius*. Invasion occurs at the point of an injury caused by insect or bird feeding, mechanical damage, growth cracks, or lesions resulting from powdery mildew. Damaged berries are quickly colonized by other fungi and bacteria. Once a single berry becomes infected, its juice begins to drip into the cluster. The juice contains fungal spores and bacterial cells that spread infection to adjacent, healthy berries. The resulting rot becomes increasingly severe as it progresses beyond the original injury. Vineyard managers need to monitor for rotting clusters by conducting visual inspections between veraison and harvest.

Sour rot management. Cultural controls for sour rot include those described for Botrytis bunch rot, including leaf removal at berry set, which increases airflow and sunlight exposure and decreases humidity around the clusters (Stapleton et al. 1990). Leaf removal has provided equivalent control to any chemical treatment tested to date. However, recent studies in table grapes did not find leaf removal to be an economical control method (Rooney-Latham et al. 2009). In addition, irrigation management is important. Over-irrigation can contribute to increased berry size and tight clusters, making the berries more prone to splitting. Plant growth

Figure 8.9. (A) Summer bunch rot (*Aspergillus niger*) and (B) sour rot on tight-clustered Zinfandel, Madera, California. *Photos* by J. C. Broome.

regulators such as GA have reduced sour rot and Botrytis bunch rot, but because it is hard to time the application correctly and use the appropriate rate, they can result in yield losses. Also, growth regulators are only allowed in certain counties, and then only with a special local needs permit.

There are some promising biological controls available for use as antagonists against the bunch rot complex. The bacterium *Pseudomonas fluorescens* has performed well in this manner (R. A. Duncan, unpublished), but it is not registered for use on grapes. Copper compounds are used as chemical control agents for sour rot. One cultural control method, removal of affected berries before they could ooze juices onto adjacent berries, decreased sour rot to levels in table grapes below those achieved by any other treatment (Rooney-Latham et al. 2009).

Downy Mildew *(Plasmopara viticola)*

Downy mildew *(Plasmopara viticola)* is a fungal disease common in areas with high summer rainfall, such as the eastern United States and Europe (Pearson and Goheen 1990), but it was unknown in California until 1990. Then in 1995 and 1998, the disease was a problem in several southern San Joaquin Valley vineyards when springtime conditions were exceptionally wet. It has been found in the Sacramento Valley, the southern San Joaquin Valley, Sierra Foothills vineyards, and nurseries in the San Joaquin Valley and Monterey County.

Symptoms, pathogen biology, and disease cycle. In the spring, yellow, oily-looking lesions develop on the upper side of the leaves, and the fungus sporulates in a dense, white-gray, fluffy growth in the lesions on the lower leaf surface (Figure 8.10). Severely infected berries and clusters may shrivel completely within days. Lesions can be yellowish and appear oily on the upper leaf surface or angular and yellow to reddish brown, depending on the age of the leaf and the lesion. Infected shoot tips thicken, curl, and become white with sporulation, eventually dying. The pathogen attacks all green parts of the vine, including leaves, fruit, shoot

Figure 8.10. Downy mildew on grapes; note oil spot symptoms and sporulation on the undersides of leaves. *Photo* by W. D. Gubler.

terminals, and tendrils. The fungus overwinters as sporangia in leaf litter and soil. Spring rains splash the zoospores onto green tissue. Young berries are more susceptible to the disease than older berries.

Downy mildew management. Use of disease-free planting materials reduces the chances of introducing downy mildew into a new vineyard. Be on the lookout for symptoms of the disease, especially when springtime is particularly wet. Eradication treatments are available for application when symptoms first appear, but these are not registered for organic production. Coppers, copper hydroxide, and basic copper sulfate plus lime (Bordeaux mixture) are organically approved alternatives. Apply fungicides starting at 10 inches of shoot growth during rainy weather and continue the applications until bloom at 10-day intervals using label rates. The fungus is generally inactive during summer, but may become active again in the fall.

Phomopsis Cane and Leafspot (Phomopsis viticola)

Phomopsis cane and leafspot, caused by *Phomopsis viticola*, is a fungal disease that is most severe when spring rainfall is prevalent. It is a common problem in northern grape-growing regions where spring rains are common after budbreak. Splashing rain is required for infection. Basal leaves that are heavily infected become distorted and usually fail to develop to full size. Shoots may be stunted or may break off at the base (although this is highly unusual in California vineyards). Severe infections may cause clusters to shrivel and dry up (also unusual in California).

Symptoms, pathogen biology, and disease cycle. Tiny dark brown to black spots with yellowish margins develop on leaf blades and veins, first appearing several weeks after a rain. This leaf blade infection distorts the leaves and turns them yellow. Black, elongated lesions appear on the base of shoots. Infected canes have a bleached appearance during the dormant season. Severely affected canes or spurs exhibit an irregular, dark brown to black discoloration intermixed with whitish, bleached areas, and within the bleached areas it is easy to see tiny, black pycnidia (the overwintering form of the fungus) from which spores arise during spring rainfalls (Figure 8.11). Infections occur when spores are rain-splashed onto green leaf or shoot tissue. The disease can generally be observed in a vineyard when rainfall occurs after budbreak in the spring.

Phomopsis management. Look for "bleached" canes to determine the likely presence of overwintering inoculum. Some growers have attempted to control the disease by pruning out infected canes or spurs in winter, but their efforts have been unsuccessful. In all areas, spring foliar treatments may be advisable if there is a high risk of rain after budbreak or if you use overhead watering for frost protection. Apply dry flowable micronized sulfur at the highest dosage permitted before the first rain and after budbreak, but before the shoots grow more than 0.5 inch long, and then again when shoots are 5 to 6 inches long or just before the next rainfall event. You can use lime sulfur as a dormant treatment at a rate of 15 gallons per acre with nozzles directed at the cordons or canes (Gubler and Leavitt, unpublished).

TRUNK PATHOGENS AND THEIR DISEASES

Eutypa dieback (Eutypa lata, E. leptoplaca)

Eutypa dieback is caused by the fungus *Eutypa lata* (Gubler and Leavitt 1992) and to a lesser degree by a related, newly discovered fungus, *E. leptoplaca* (Trouillas and Gubler 2004). The disease develops over several years, but first symptoms are evident 2 to 3 years after initial infection. Eutypa dieback tends to occur in older vineyards and generally after large cuts have been made on the vine. The pathogens invade the vine's water-conducting tissue through pruning wounds and move both acropetally and basipetally in the cordon. The fungus forms

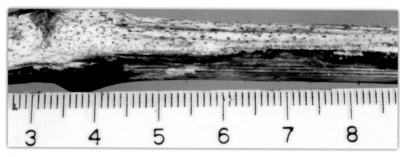

Figure 8.11. Pycnidia and black streaking of Phomopsis viticola on a bleached, dormant cane. Scale is in centimeters. *Photo* by J. C. Broome.

wedge-shaped cankers in the permanent wood of the vine and over time causes death of spurs, cordons, and, ultimately, the entire vine.

Symptoms, pathogen biology, and disease cycle. Eutypa causes characteristic foliar symptoms on grapevines, including chlorosis and tattering of the leaves and stunting of the shoots (Figure 8.12). Symptoms in the wood are characterized by wedge-shaped, brown cankers that develop in the vascular tissue. The fungus produces sexual spores (ascospores) in perithecia on dead wood. The perithecium is a specialized fruiting body embedded in a blackened substrate called a stroma. This structure develops in regions with more than 16 inches of annual rainfall and has been found on 21 different hosts in the North Coast and Delta production areas (Figure 8.13). The fungus also forms asexual spores (spermatia), which form in another kind of fruiting body, a pycnidium, on standing grapevine wood. However, the spermatia do not appear to play any role in disease development.

Eutypa is spread to new pruning wounds by wind-driven and water-splashed ascospores that are released and spread during rain events in the fall and winter. Having entered the pruning wounds, the fungus invades the xylem tissue, weakens the vine by producing toxins, and gives off cell wall-degrading enzymes that kill the wood (Rolshausen, Trouillas, and Gubler 2004). The fungus moves in both directions—outward toward the tip of the cordon (acropetally) and downward into the trunk (basipetally)—in a vine and can take 2 to 4 years before it expresses symptoms, depending on the grapevine variety. The fungus overwinters in stroma on numerous different hosts in California, including grapes, apricot, Ceanothus (Moller and Kasimatis 1978), cherry (Munkvold and Marois 1994), kiwifruit, blueberry, pear, California buckeye, oleander, big leaf maple, and willow, among others (Trouillas and Gubler 2004).

Eutypa management. Monitoring for Eutypa foliar symptoms on grapevines should be conducted in late spring, when characteristic leaf chlorosis and tattering and stunted shoots can be observed and

Figure 8.12. Disease cycle of Eutypa canker disease *(Eutypa lata, E. leptoplaca)*: (A) stroma on a dead trunk or alternate hosts; (B) ascospores released; (C) fresh pruning wounds; (D) several years after infection; (E) canker. *Photos* by F. Trouillas and W. D. Gubler.

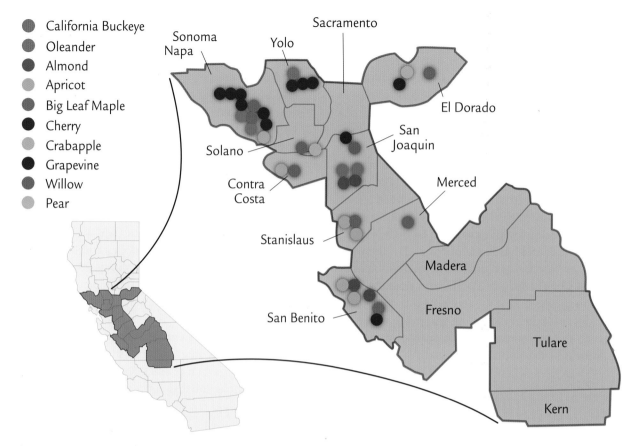

Figure 8.13. Geographical distribution and hosts of *Eutypa lata* in California, sampling conducted from 2000–2002. Adapted from Gubler et al. 2005b.

Legend:
- California Buckeye
- Oleander
- Almond
- Apricot
- Big Leaf Maple
- Cherry
- Crabapple
- Grapevine
- Willow
- Pear

Map labels: Sonoma Napa, Yolo, Sacramento, El Dorado, San Joaquin, Solano, Merced, Contra Costa, Stanislaus, Madera, San Benito, Fresno, Tulare, Kern

before healthy shoot growth has a chance to cover the symptomatic leaves. To manage the disease, keep the number and size of pruning cuts to a minimum and train vines properly to avoid the need for large cuts later in retraining. Prune out infected spurs and cordons as late in the dormant period as possible, after rains have reduced the spore load and when the vine will heal most quickly (Munkvold and Marois 1995). Be aware, however, that late pruning can lead to delayed budbreak for some varieties in some areas, a factor that you should take into consideration when you schedule your pruning. Pruning wounds remain susceptible for 4 to 6 weeks if made in December, but only for about 7 to 10 days if made in March (Moller and Kasimatis 1980). Retraining is one technique for reducing Eutypa dieback in vines. To do it, you cut out dead sections of cordons and retrain a cane from an uninfected part of the vine to fill the space. Double pruning is becoming more and more common. It involves using a mechanized

pruner once in the fall or winter to leave canes of 10 to 12 buds or more and then hand pruning vines in the late-dormant period (Weber et al. 2007)

Recent work suggests that local sources of inoculum on native species like *Ceanothus*, willow, and bay laurel are more important than was formerly thought, so careful sanitation or removal of stroma on dead grapevines and other species in the vicinity of a vineyard are probably appropriate as well (Gubler et al. 2005a). However, it is thought today that most inoculum comes from within the infected vineyard and care should be taken to remove diseased wood and burn it or remove it from the vineyard area.

Pruning wounds are colonized by a variety of fungi, such as *Cladosporium* and *Fusarium* species, as well as others that can then exclude the pathogen from invading the wounds and causing disease (Munkvold and Marois 1993). Biological control agents have been shown in tests to protect pruning wounds, but no products are currently registered for this use on organic grapes in California.

Chemical treatments are most effective if applied directly to the pruning wounds immediately after pruning. Sulfur, oils, coppers, and *Bacillus subtilis* were not found to be effective against *Eutypa*. Boric acid at 5,000 ppm and one synthetic detergent are highly effective for *Eutypa* control as well as for control of the pathogens that cause black measles (Rolshausen and Gubler 2005).

Bot Canker of Grapevines (*Botryosphaeriaceae* fungi)

Bot canker caused by *Lasiodiplodia theobromae* (previously *Botryrosphaeria theobromae*) was once thought to be the main cause of cordon death and dieback in the southern San Joaquin Valley region (Gubler and Leavitt 1992). However, more recent studies (Úrbez-Torres et al. 2006; Úrbez-Torres, Luque, and Gubler 2007) have shown certain closely related fungi in the family Botryosphaeriaceae to be the most prevalent fungal species associated with grapevine cankers in California. Like *Eutypa*, these fungi enter the vine through pruning wounds and move in both directions in the cordons, killing tissue as they go. Currently, nine fungal species in the family Botryosphaeriaceae have been found to be associated with grapevine cankers in California. All nine species have been proven to be pathogenic on grapevines, and some of them have been shown to be able to colonize the wood much more quickly than *E. lata,* and thus to kill spurs and cordons much more quickly (Úrbez-Torres and Gubler 2009).

The fungi form cankers in the permanent wood of the vine, over time causing the death of spurs, cordons, and, ultimately, the entire vine.

Symptoms, pathogen biology, and disease cycle. Visible symptoms in the wood of vines cankered by all of these fungi are similar, characterized as they are by wedge-shaped, brown cankers that develop in the vascular tissue (Figure 8.14). In addition, some members of the Botryosphaeriaceae (such as *Neofusicoccum parvum* and *Botryosphaeria dothidea*) have been shown to cause dark streaking of the wood (Úrbez-Torres and Gubler 2009). However, unlike Eutypa, the cankers caused by Botryosphaeriaceous fungi have not yet been shown to be associated with any foliar symptoms in California. These fungi produce fruiting bodies (pycnidia) on the surface of the canker or on old, dead wood on cordons and spurs. The pycnidia produce spores that are dispersed with winter rains. Recent spore-trapping studies conducted throughout nine different grape-growing regions in California showed spores of Botryosphaeriaceae to be discharged from the first rain of the early fall until the last rain of the spring. However, the highest numbers of spores were trapped following rainfall during the winter months of December, January, and February (Úrbez-Torres and Gubler 2008a). Botryosphaeriaceae spore release was also observed to be triggered by overhead sprinkler irrigation in California vineyards. Spores land on newly cut pruning wounds. Cankers or wood streaking caused by the Botryosphaeriaceae are most frequently seen in vineyards that have been established for 10 or more years. However, a recent

Figure 8.14. Bot canker (Botryosphaeriaceae fungi) of grapevines: Picnidia or teleomorphs (A) on vine and (B) in close-up; (C) micrograph of the pathogen. *Photos* by J. Urbez.

study conducted in Mexico showed cankers caused by *Lasiodiplodia theobromae* to be present in vines only 5 or 6 years old (Úrbez-Torres et al. 2008).

Management of bot canker. Important cultural controls include training and pruning practices. The focus should be on minimizing the amount of inoculum that can enter the vine by reducing opportunities, both in terms of space and time. Spatially, you need to minimize the number and size of vines and train the vines properly from the start, to avoid the need for large cuts that are required for retraining. Prune as late in the dormant period as possible, when rains have reduced the spore load and the vine will heal most quickly. Pruning out dead sections of cordons and retraining a cane from an uninfected part of the cordon to fill the resulting gap is a good way to stop disease. Chemical treatments are most effective if applied directly to the pruning wounds immediately after pruning and are the same as those listed for Eutypa, above. Double pruning, a cultural method shown to reduce infections caused by *E. lata,* has been also shown to effectively reduce infections caused by Botryosphaeriaceae species (Úrbez-Torres and Gubler 2008b).

OTHER FUNGAL CANKER DISEASES

Other canker disease pathogens have also been identified in California (Trouillas and Gubler, unpublished). These include species of *Cryptovalsa, Cryptosphaearia, Diatrype,* and *Diatrypella.* These pathogens belong to the same family of fungi as *Eutypa* and cause symptoms very similar to those caused by *Eutypa.* However, sensitive molecular tests called polymerase chain reaction (PCR) have been used to successfully differentiate these pathogens. The disease cycles appear to be similar to those of *Eutypa* as well, with infection being initiated by spores that have been deposited onto pruning wounds by wind and rains. There are many other hosts of these fungi in California, and the pathogens themselves are not well understood. Control methods are similar to those used for Eutypa: late pruning, double pruning, and protective treatments of pruning wounds.

Esca (black measles) and Petri disease (young esca)

Esca (black measles) and vine declines including young esca, or Petri disease, are caused by a complex of fungal pathogens. Esca typically occurs in older grapevines and is caused primarily by *Phaeoacremonium aleophilum (Togninia minima)* as well as *Phaeomoniella chlamydospora.* Petri disease, also known as young esca, occurs on immature grapevines and is typically caused when the roots are infected by the fungus *Phaeomoniella chlamydospora.* However, both of these fungi can be found in plants affected by both disorders. When diseased wood is obtained from nurseries or disease occurs in newly planted vineyards, you can see symptoms in vines as young as 2 or 3 years old (Khan et al. 2000).

Symptoms, pathogen biology, and disease cycle. Leaf symptoms of esca include small chlorotic interveinal areas that develop into a "tiger stripe" pattern that enlarges and dries out (Figure 8.15). In red varieties, those areas are surrounded by dark red margins. Severely affected vines will exhibit leaf drop and cane dieback. The fruit can have small, dark spots that may be surrounded by a purple ring. In severely affected fruit, the berries will crack and dry out or will raisin or wilt (Gubler et al. 2004). Scheck et al. (1998) isolated the causal fungus from vines and identified it as *Phaeoacremonium* spp. Recent work has shown in greater detail that there is a group of fungi that occur as weak endophytic pathogens, with *Togninia minima* being the most prevalent species. These fungi, all belonging to the genus *Phaeoacremonium,* produce perithecia in old, rotted, vascular tissue of pruning wounds and in cracks in cordons, trunks, and spurs (Rooney-Latham, Eskalen, and Gubler 2005a). Spores are released with rainfall and the sexual spores (ascospores) re-infect the grapevine through pruning wounds, which remain susceptible for up to 16 weeks (Eskalen and Gubler 2002; Rooney, Eskalen, and Gubler 2002; Rooney-Latham, Eskalen, and Gubler 2005a). Insect transmission of sexual spores may also occur (Eskalen et al., unpublished). The pathogen overwinters as perithecia or in the endophytic phase.

Symptoms of Petri disease (young vine decline or young esca) include vascular streaking of the woody cylinder, stunted growth, small chlorotic leaves, and shoot tip dieback (Figure 8.15). *Phaeomoniella chlamydospora* resides as pycnidia on grapevines in 3- to 5-year-old pruning and girdling wounds, and its spores are released with rainfall during the months of November through April (Eskalen and Gubler 2002; Rooney-Latham, Eskalen, and Gubler 2005b). Symptoms generally are expressed either in the year that new infections occur or 1 year later.

New diagnostic methods include use of PCR tests to distinguish between nine species of *Phaeoacremonium* and the single species of *Phaeomoniella (chlamydospora)* (Eskalen, Gubler, and Khan 2001). Several other fungi have also been associated with internal esca symptoms, including *Phialophora* and *Caudophora* spp. (Gubler et al., unpublished).

Esca management. It is clear that pruning time is not as important for this disease as it is for Eutypa dieback or bot cankers (Figure 8.15), because esca pathogens are capable of infecting pruning wounds made at any time during the winter and spring. Double pruning probably would work best for spur-pruned vines (Weber, Trouillas, and Gubler 2007), but larger wounds must be protected. Chemicals that are used for *Eutypa* are also effective against esca pathogens.

Petri disease or young esca management. It is important to obtain clean, healthy planting stock, to plant using appropriate cultural practices, and to provide sufficient irrigation and fertilization to young, newly planted vines if you want to avoid establishment problems (Rooney et al. 2002). These pathogens, which are common soil inhabitants, also live as epiphytes on the foliage, fruit, and bark of grapevines and as endophytes in the water-conducting tissue of grapevines. If the vines are stressed, the pathogens appear to become more virulent and to cause disease by producing toxins that injure the plant tissue (Figure 8.15). Research has shown that types of stress that predispose a grapevine toward Petri disease include early fruiting of the vines (before year three), "J" rooting, and poor irrigation management or water deficit stress.

Once the stress has occurred and the vines display symptoms, you must remove them and replant with new vines. You do not have to remove the entire vineyard, only the vines that show symptoms. Grapevine varieties and rootstocks vary in their susceptibility. All of the phylloxera- and nematode-resistant stocks are more susceptible than AXR1 rootstock (Feliciano, Eskalen, and Gubler 2004). For example, the rootstocks 3309, 101–14, and 5C are very susceptible to *P. chlamydospora,* whereas AXR1 is nearly immune to the disease. Hot water dips in the nursery have been shown in some research to reduce disease levels to the point where control can be achieved. Work has shown that temperatures of 124°F (51°C) will shock the pathogen in wood but not necessarily kill the fungus (Whiting, Khan, and Gubler 2001).

Black foot *(Cylindrocarpon destructans, C. liriodendron)*

Black foot is a root disease caused independently by two distinct fungi: *Cylindrocarpon destructans* and *C. liriodendron.* This disease is relatively new to California vineyards. Its first occurrence here coincided with the development and use of a rootstock that was bred to be resistant to phylloxera and nematodes. Black foot is a juvenile plant disease that is more prevalent in heavy, poorly drained soils. As grapevines mature they become less susceptible to infection and the disease is less prevalent. The soilborne pathogens infect small feeder roots and then invade larger roots and work their way into the tap root and upward into the upper tap root of the vine. Once it has invaded this tissue, the disease will kill the vine.

Figure 8.15. Esca (black measles) and young esca (vine decline): (A) teleomorphs in dead wood, vascular tissue; (B) vine showing signs of stress and decline; (C) berry symptoms. *Photos* by W. D. Gubler.

Symptoms, biology, and disease cycle. Symptoms of black foot are various. On roots, the fungus causes small, black lesions on feeder and lateral roots. These lesions are hard and dry and therefore easy to distinguish from phytophthora lesions, which are wet and slimy. Internal symptoms include a diffuse blackening of the vascular tissue around the pith. As the fungus grows through the xylem, the wood dries and takes on a purplish silver color. Foliar symptoms include water-stressed leaves with a dull appearance and dead blotches. Stunting of shoots occurs, and death of the vine can follow soon after.

Cylindrocarpon spp. live in the soil and invade the roots of susceptible hosts. This disease was unknown in California until the rootstock AXR1 was phased out due to its susceptibility to phylloxera. Unlike AXR1, the new phylloxera-resistant rootstocks have been shown to be very susceptible to the disease. The Black foot pathogens can move with nursery stock, so new vines from nurseries may already be infected.

Black foot management. Purchase clean material from nurseries. Check that it is clean by cutting into several representative vines before you plant to check the xylem at the base of the vine for internal, diffuse black discoloration around the pith. Discard any vines that show these symptoms. Also, plant your vines in well-drained soil and do not overirrigate.

MAJOR BACTERIAL DISEASES

Pierce's Disease *(Xylella fastidiosa)*

Pierce's disease is caused by the bacterium *Xylella fastidiosa*, which lives in the water-conducting tissue of plants (xylem) and is spread from plant to plant by xylem-feeding sharpshooters (Hopkins 1989; Redak et al. 2004; Purcell 2005). Pierce's disease appears to be restricted to portions of North America with mild winters. It is less prevalent where winter temperatures are colder, such as at higher altitudes and farther inland, away from marine influences. At more northern latitudes (above the 44th parallel, about 100 miles north of the California-Oregon border), the disease does not occur. Within California, the disease was historically found near riparian areas, due to the natural habitat of the native vectors. With the introduction of a new vector, the glassy-winged sharpshooter (*Homalodisca vitripennis*), however, it has now become a more widespread and more serious threat.

Symptoms, pathogen biology, and disease cycle. Symptoms of Pierce's disease first appear as water stress in midsummer and are caused by bacteria blocking the vine's water-conducting tissue. Host plant resistance products and toxins also contribute to xylem blockage and thus to water stress and plant symptoms. Leaves become slightly yellow or red along the margins in white and red varieties, respectively, and eventually the leaf margins dry out or die in concentric zones. The dried leaves fall, leaving the petiole (leaf stem) attached to the cane. By midseason, some or all fruit clusters on infected canes may wilt and dry and the tips of canes may die back. Vines may deteriorate rapidly after symptoms appear. Wood on new canes matures irregularly, producing patches of green surrounded by mature brown bark. In the year of its initial infection, a vine will usually show Pierce's disease symptoms on only one or two canes late in the season. Symptoms gradually spread along the cane from the point of infection outward toward the end and more slowly toward the base. The following year will see some canes or spurs failing to bud out. From late April through summer, infected vines may grow at a normal rate, but the total new growth is less than that of a healthy vine. Not all vines that are infected will develop the disease. The probability of recovery depends on the variety, the date of infection, and the age of the vineyard (Varela, Smith, and Phillips 2001). Once the vine has been infected for more than a year (i.e., bacteria have survived the first winter), recovery is much less likely. Young vines are more susceptible than mature vines, probably because much less wood is pruned from vines during the training period than when they are mature (Varela, Smith, and Phillips 2001).

Native sharpshooter vectors are active in the spring after average temperatures warm up above 59°F and they can transmit the bacterium to the vines anytime thereafter. The blue-green sharpshooter (*Graphocephala atropunctata*) is the most important vector in coastal areas. Its principal breeding habitat is riparian (riverbank) vegetation, although ornamental landscape plants may also harbor breeding populations. As the season progresses, the insects shift their feeding preference toward plants with succulent growth. The green sharpshooter (*Draeculacephala minerva*) and the redheaded sharpshooter (*Carneocephala fulgida*) are also present in coastal areas, but they are more important as vectors of this disease in the Central Valley. In the Central Valley, irrigated pastures, hay fields, and grasses on ditch backs are the principal breeding and feeding habitats for green and redheaded sharpshooters. These two grass-feeding sharpshooters also occur along ditches, streams, and roadsides where grasses and sedges provide suitable breeding habitat (Varela, Smith, and Phillips 2001).

The glassy-winged sharpshooter (GWSS) has recently become established in various parts of California, principally the South Coast and southern San Joaquin Valley. This new vector, though a less efficient vector of the pathogen than native sharpshooters, is a serious threat to California vineyards because it moves into vineyards more quickly and over greater distances than the other sharpshooter species, and it feeds on mature grapevines, leaves, and dormant vines (Redak et al. 2004; Blua, Phillips, and Redak 1999). The GWSS occurs in unusually high numbers in citrus and avocado groves and on some woody ornamentals. It has become locally abundant in Riverside and San Diego Counties. The GWSS is expected to spread north and become a permanent resident of various habitats throughout northern California. Crape myrtle and sumac are especially preferred hosts. It reproduces on eucalyptus, coastal live oaks, and a wide range of trees in southern California. Because the GWSS is a relatively new arrival to the state, it is not clear yet where in the state and in which habitats it will become permanently established.

Pierce's disease management. Monitor for the blue-green and glassy-winged sharpshooters with yellow sticky traps and tape as well as sweep nets. Use a sweep net on neighboring pastures and ditches for the green and redheaded sharpshooters, as they are not attracted to yellow sticky traps. Make visual observations for vectors and symptoms of Pierce's disease on existing vines as well as on neighboring vegetation, making a note of when symptoms develop (Varela, Smith, and Phillips 2001).

There are no effective control measures currently available for Pierce's disease when infected sharpshooter vectors are present. In general, a grower should take steps to reduce the sharpshooter population, reduce the bacterial population in neighboring vegetation, and eliminate pathogen sources within the vineyard by removing diseased vines. For more information on vector control for Pierce's disease, see chapter 9.

Infections that appear in the spring and summer can be removed with fall pruning. Severe pruning can eliminate Pierce's disease from infected vines: in fall and winter, prune back infected vines to within a few inches of the graft union and then retrain a new shoot to fill the void the next spring. Vines that have had Pierce's disease symptoms for more than 1 year should be considered a source of infection and removed.

Some infected vines recover from Pierce's disease after their first winter, with the likelihood of recovery depending on the date of infection. Early infections (up until June) appear to be most likely to survive until the following year. Late infections (after June) by blue-green sharpshooters, green sharpshooters, and redheaded sharpshooters are least likely to persist the following growing season. This is not the case with the GWSS, however, because it feeds on leaves near the base of the cane. Vine recovery rates depend on grape variety and susceptibility: recovery is higher in Chenin Blanc, Sylvaner, Ruby Cabernet, Zinfandel, Thompson Seedless, and White Riesling as compared to Barbera, Chardonnay, and Pinot Noir. Cabernet Sauvignon, Grey Riesling, Merlot, Napa Gamay, Petite Sirah, and Sauvignon Blanc are intermediate in their susceptibility to this disease, and thus, in their probability of recovery (Table 8.1) (Varela, Smith, and Phillips 2001; Purcell 2005). However, any vine that has been infected for more than a year (i.e., the bacteria have survived their first winter) is much less likely to recover.

MAJOR VIRUS DISEASES

Grapevine viruses can cause debilitating diseases that tend to have characteristic symptoms. Viruses are tiny, replicating pieces of genetic material (single- or double-stranded DNA or RNA) covered with a protein coat. Viruses are much smaller than fungi or bacteria and require specialized testing for detection, which ranges from inoculation of specific indicator plant species to serological tests such as the enzyme-linked immunosorbent assay (ELISA) or highly sensitive DNA tests such as reverse transcriptase polymerase chain reaction or real-time polymerase chain reaction (RT PCR). It is important to recognize the various diseases that are caused by viruses and to avoid using infected vines for propagation or when planting new vineyards.

Grapevine virus diseases can be spread by rootstock or scion cuttings from infected plants or by vectors such as nematodes and some sucking insects. The best management approach (and often the only one) for viruses is to prevent them in the first place and use clean, certified planting stock from a reputable nursery. One of the most common ways in which virus is spread in California is through field budding. Growers often obtain scion wood from commercial vineyards that are thought to be of superior quality. When the source vineyard has a virus, however, it is spread to the new vineyard through use of the infected wood.

Clean wood of the desired selection may be scarce at nurseries and more expensive. While these considerations are important, as a grower you should do everything possible to avoid spreading virus with scion wood. Visit the vineyards you plan to graft from often in the growing season and identify high-yielding, non-symptomatic grapevines from which to harvest dormant canes for propagation. Remember, though, that a lack of symptoms in the source vineyard is no guarantee that there is no virus present. Many of the major grapevine viruses are latent, meaning that they do not show symptoms during some or all of the growing season. Selected grapevines should therefore also be pre-tested for virus by a competent diagnostic laboratory.

There are four main virus diseases that are of concern to winegrape growers and will be covered here briefly; many others occur fairly infrequently and will not be covered. More detailed information on grapevine virus diseases is available from other sources (Golino, Rowhani, and Uyemoto. in press).

Fanleaf Degeneration

Fanleaf degeneration is considered a fairly major viticultural problem throughout most of the wine-growing regions in California, as well as around the world. It is also the oldest known virus disease of *V. vinifera,* possibly present since the earliest known cultivation in the Mediterranean, and it has been found in European herbarium samples. Leaves, canes, and shoots can be affected. Symptoms include the open sinuses at the base of the leaves that give it the "fan leaf" name, bright yellow vein banding along major veins, and leaf yellowing (Figure 8.16). Leaf symptoms start to appear in the spring and continue into early to midsummer. Disease severity will vary with variety, rootstock, and environmental conditions, and also from one year to the next. Although it rarely kills a vine, fanleaf degeneration does cause them to become unfruitful by producing shot berries (Figure 8.16).

The disease is vectored by the dagger nematode, *Xiphinema index.* The nematode initially becomes infected when it feeds on diseased vines. Adults and juveniles can transmit the virus, which can stay in the nematode vector for up to 8 months in the absence of host plants, or for up to 3 months when the nematode feeds on virus-immune host plants (Pearson and Goheen 1990). Infected living roots left in the soil after vine removal can serve as reservoirs for infection of newly planted vines, and many years (as many as 10) of fallowing are needed before a healthy new vineyard can be established. The causal virus, *Grapevine fanleaf virus* (GFLV), is a member of the nepovirus group composed of single-stranded RNA in two functional entities, both of which are needed for infection to occur. You can detect the presence of the virus through transmission to herbaceous indicator species, grafting to the woody

Figure 8.16. Grapevine fanleaf virus chrome mosaic symptoms include (A) bright yellow leaves with shadow-look effect and (B) shot berries, also known as "hens and chicks." *Photos* by S. T. Sim.

indicator *V. rupestris* St. George in a 2-year field test, ELISA testing, or by RT PCR testing. Spring is the ideal time to test for GFLV (Golino, Rowhani, and Uyemoto, in press).

The use of clean rootstock and scions is the single most important tactic for avoiding virus problems. Once the virus and vector are present in a vineyard, replantings will become infected. Therefore, only long fallows (up to 10 years), long enough to allow all old infected roots to die out, are available as a control measure. In addition, recent breeding for nematode and virus resistance in rootstocks has produced 039-16, a hybrid of *V. vinifera* and *V. rotundifolia* that has demonstrated some field tolerance to both *X. index* and *fanleaf degeneration virus*. The 039-16 hybrid has *V. vinifera* parentage, however, and therefore a possible link to phylloxera susceptibility, so it is only recommended for sites that are known to be infested with nematode-virus complex (Golino, Rowhani, and Uyemoto, in press).

Leafroll

Leafroll occurs worldwide wherever grapes are grown and is common throughout California. It is caused by several different *Grapevine leafroll-associated viruses* (GLRaV). Symptoms develop in early summer in water-stressed vineyards and later in the season in well-irrigated vineyards. On vines with leafroll, the margins of the leaf blades roll downward and areas between major veins turn yellow (on green-grape varieties) or red (on red-grape varieties) (Figures 8.17 and 8.18). The area around the major veins on both green and red varieties remains distinctly green. The disease does not produce obvious symptoms in the dormant or early growing seasons, except for infected vines being slightly smaller than healthy vines. Rootstocks do not show any symptoms during any time of the year, due to their American parentage. However, the virus can move from infected scion wood to rootstocks and from infected rootstocks to scions.

As of this writing, there are nine recognized, serologically distinct viruses associated with grapevine leafroll disease. Taxonomically, the nine GLRaVs are classified in the virus family Closteroviridae, which is characterized by large, flexuous rod-shaped particles that range in length from 1,250 to 2,200 nm. The nine viruses are *Grapevine leafroll associated virus* numbers 1 through 9 (abbreviated as GLRaV-1, -2, etc.). Phylogenetically, GLRaVs-1, -3, -5, and -9 are in the genus *Ampelovirus*, whose members are vectored by mealybugs. One or more mealybug species have been shown experimentally to be vectors of GLRaVs-1, -3, -5, and 9. Although classified similarly in genus *Ampelovirus*, GLRaV-4 currently lacks evidence of vector transmission. GLRaV-2 is in the genus *Closterovirus* (closteroviruses are vectored by aphids). Not enough is known about GLRaVs-6, -7, or -8 for group classification. In California, biological indices for leafroll are done on the woody indicator Cabernet Franc and involve grafting chip buds from the accessions under test (the candidate grapevine) onto potted plants (Golino, Rowhani, and Uyemoto, in press).

Primary sources of the disease's spread include propagation materials, and its secondary spread is via several mealybug species and soft scale insect vectors. To date, five species of mealybugs found in California—obscure, longtailed, citrus, grape,

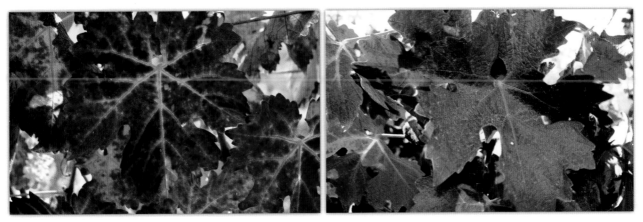

Figure 8.17. (A) Typical leafroll symptoms on Cabernet Sauvignon grafted to the rootstock St. George and (B) atypical leafroll symptoms on Cabernet Sauvignon grafted to the rootstock 3309C. *Photos* by S. T. Sim.

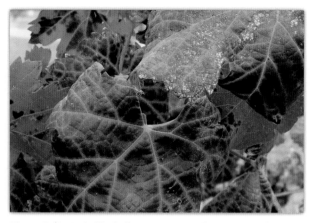

Figure 8.18. Typical leafroll symptoms on Cabernet Franc, showing interveinal burgundy red coloration, green veins, and downward rolling of the leaf margin. Symptoms appear in the fall on the basal leaves. *Photo* by S. T. Sim.

and vine mealybugs—have been found to vector GLRaV-3, a common virus detected frequently in California. Several mealybug species can also vector GLRaVs-1, -5 and -9 (Golino, Rowhani, and Uyemoto, in press).

Virus management methods include the use of clean propagation materials (scions and rootstocks), removal of infected vines, and avoiding the introduction of vectors into a vineyard if possible. The use of pesticides to control vectors is difficult, particularly in an organic vineyard, and it is not thought to be efficacious.

Corky Bark

Corky bark occurs worldwide where grapes are grown. It is less common, however, than leafroll. Corky bark disease can stay latent in many cuttings, only to show up later when infected buds are grafted onto phylloxera-resistant rootstocks. When this occurs, an incompatibility reaction gradually develops between the rootstock and the scion, causing the scion to slowly decline and die. Corky bark symptoms are similar to those of leafroll, but vine vigor is more severely reduced. After several years the scion will die, leaving only the infected but symptomless rootstock. Do not use that rootstock for new planting material.

The disease causes reduced shoot growth in spring, and the leaves on infected vines turn yellow by summer, later rolling downward and turning red to bronze-ish yellow. Unlike leafroll, the changes in leaf color occur across the entire leaf, including the major veins. In fall, the leaves do not drop off as usual, but remain on the vine long after the first frost. Yield and quality are reduced. Some rootstocks show no symptoms, while others develop deep pits and grooves in the stem near the graft union.

Corky bark disease is associated with a vitivirus, *Grapevine virus B* (GVB). It can be transmitted to healthy grapevines by the longtailed mealybug, *Pseudococcus longispinus*. To detect the virus, researchers use as an indicator a grapevine rootstock hybrid LN 33, which is a cross of Couderc 1613 *V. vinifera* cv. Thompson Seedless. In the second year after grafting with infected wood, LN 33 will predictably develop symptoms of corky bark disease, including grooves and pits on the woody cylinder, bark splitting, and the swelling of canes and proliferation of spongy callus tissue (Golino, Rowhani, and Uyemoto, in press). The primary method of control is to prevent the problem entirely by planting only healthy scion material and rootstocks.

Rupestris Stem Pitting

Rupestris stem pitting occurs in vines imported to California from Western Europe after 1950. It causes a slow decline in vines that typically becomes apparent only 4 years after planting, when yields are reduced and maturity delayed. Specific symptoms on wood and leaves do not appear on most scions or rootstocks, except that the leaves on infected vines do tend to be smaller and the canes tend to be smaller in diameter. There is no leaf color change related to this virus, but fruit clusters on affected vines are smaller and their ripening is delayed. On St. George and a few other rootstocks derived from *V. rupestris* parentage, you can remove bark from the

Figure 8.19. Grapevine trunks after the bark was peeled off to show the woody cylinder. Two trunks on left show pitting and grooving as a result of a virus infection of the rugose wood family of viruses. Healthy trunk is at right. *Photo* by S. T. Sim.

rootstock to reveal that its wood cylinder appears to be covered with small pits (Figure 8.19).

Rupestris stem pitting-associated virus (RSPaV) is the causal agent of this disease. It is normally of little consequence, although some reports suggest it causes reduced vigor or decline in vines grafted onto rootstocks with *V. rupestris* parentage. RSPaV is widely distributed in the world's vineyards and is allowed in most grapevine certification programs. The California Department of Food and Agriculture (CDFA) grapevine Registration and Certification (R&C) program allows the virus to be present in certified stock, but efforts are under way to eliminate it from the program (Golino, Rowhani, and Uyemoto, in press) insofar as is possible.

CONCLUSION

We have described the most important diseases that a winegrape grower might encounter in California, including diseases caused by foliar, fruit, and wood-rotting fungal pathogens, the most important bacterial pathogen (*Xyella fastidiosa*), and four key viruses. Organic winegrape growers must set up their vineyards with attention to the site's slope and aspect, trellis systems, vineyard layout, and row orientation to take full advantage of microclimate effects, which can provide some disease control. Best management practices must be followed with regard to the selection of clean planting stock and the management of soils and vines, in particular vine vigor, as described in other chapters of this manual. Despite the practices described in this chapter that should help to minimize disease, growers may still find they must take specific disease control measures, particularly for powdery mildew control, on an annual basis. Organic disease management must be integrated into the annual production cycle with emphasis on cultural practices such as pruning, canopy, and cover crop management, among others. Key actions—including monitoring, roguing of infected plant tissue, and application of permitted materials when needed—must be taken at appropriate times throughout the growing and dormant seasons (Table 8.4).

Table 8.4. Vine growth stages and seasonally based organic treatments for major diseases

| Disease | Time of season | | | | | | | | |
	Dormant	Bud-break	Pre-bloom	Bloom	Fruit set	Pre-close	Veraison	Pre-harvest	Harvest
Powdery mildew	Apply lime sulfur if high powdery mildew levels in previous season.	Wettable sulfur	Dust	Rotation of materials in Table 8.2 until veraison		Leaf removal	End of mildew treatments		
Botrytis bunch rot	Pruning for vigor control			Copper		Leaf removal		Copper	
Canker diseases	Late pruning, treat cuts		Monitor and mark symptomatic vines.						
Phomopsis	Lime sulfur		Wettable sulfur						
Pierce's disease	Severe pruning of infected vines	Monitor for vectors. Surround if vectors are present.			Assess neighboring vegetation for Pierce's disease reservoirs and take action as needed.				

General notes on cultural practices for disease control:
Leaf removal should be done by the end of cluster set. This timing increases the berries' ability to withstand sunburn and reduces dead blossom debris in clusters.
Pruning should be done as late in the dormant season as possible to manage vigor, provide crop load, and control canker diseases.
Cover crops assist with vineyard access for dormant treatments. Mow in spring to reduce risks of increased humidity, and thus mildew. Manage for weed suppression. Mow late in the season to reduce pre-harvest bunch rot risks.
Suckering is done by hand when shoots are 6–10 inches in length to thin the canopy and improve the microclimate to reduce mildew and bunch rots.
Shoot positioning, using shoot positioning wires (if present) in the trellis system improves spray coverage and improves the microclimate for control of mildew and gray mold.

REFERENCES/RESOURCES

Bledsoe, A. M., W. M. Kliewer, and J. J. Marois. 1988. Effects of timing and severity of leaf removal on yield and fruit composition of Sauvignon blanc grape vines. American Journal of Enology and Viticulture 39:49–54.

Blua, M. J., P. A. Phillips, and R. A. Redak. 1999. A new sharpshooter threatens both crops and ornamentals. California Agriculture. 53:22–25.

Broome, J. C., J. T. English, J. J. Marois, B. A. Latorre, and J. C. Aviles. 1995. Development of an infection model for Botrytis bunch rot of grapes based on wetness duration and temperature. Phytopathology 85:97–102.

California Department of Food and Agriculture (CDFA). 2005. GWSS Biological Control Newsletter, spring 2005. www.cdfa.ca.gov/gwss/BioCtrlRep/Doc/Spring2005.pdf.

Chellemi, D. O., and J. J. Marois. 1991a. Effect of fungicides and water on sporulation of Uncinula necator. Plant Disease 75(5):455–457.

———. 1991b. Development of a demographic growth model for Uncinula necator by using a microcomputer spreadsheet program. Phytopathology 81(3):250–254.

———. 1992. Influence of leaf removal, fungicide applications, and fruit maturity on incidence and severity of grape powdery mildew. American Journal of Enology and Viticulture 43(1):53–57.

Crisp, P. 2006. An evaluation of biological and abiotic controls for grapevine powdery mildew. 1. Greenhouse studies. Australian Journal of Grape Wine Research 12(3):192–202.

Dell, K. J., W. D. Gubler, R. Krueager, M. Sanger, and L. J. Bettiga. 1998. The efficacy of JMS stylet oil on grape powdery mildew and Botrytis bunch rot and effects on fermentation. American Journal of Enology and Viticulture 49:11–16.

Delp, C. J. 1954. Effect of temperature and humidity on the grape powdery mildew fungus. Phytopathology 44:615–626.

Duncan, R. 2000. Summer bunch rot trials. www.agro-k.com/research/nz-trials/2000-001.htm

Eskalen, A., and W. D. Gubler. 2002. Association of spores of Phaeomoniella chlamydospora, Phaeoacremonium inflatipes, and P. aleophilum with grapevine cordons in California. Phytopathologia Mediterranea 40:S429–432.

Eskalen, A., W. D. Gubler, and A. Khan. 2001. Rootstock susceptibility to Phaeomoniella chlamydospora and Phaeoacremonium spp. Phytopathologia Mediterranea 40:S433–438.

Feliciano, A. J., A. Eskalen, and W. D. Gubler. 2004. Differential susceptibility of three grapevine cultivars to Phaeoacremonium aleophilum and Phaeomoniella chlamydospora in California. Phytopathologia Mediterranea 43:66–69.

Gadino, Angela N. 2007. Evaluation of whey powder to control powdery mildew, Erysiphe necator (Schw.) on winegrapes. Thesis. California Polytechnic State University.

Gadoury, D. M. 1995. Controlling fungal diseases of grapevines under organic management. www.nysaes.cornell.edu/hort/faculty/pool/organicvitwkshp/tabofcontents.html.

Golino, D., A. Rowhani, and J. Uyemoto. In press. Grapevine virus diseases. In Grape pest management, 3rd ed. Oakland: University of California, Division of Agriculture and Natural Resources, Publication 3343.

Gubler, W. D., and D. J. Hirschfeldt. 1992. Powdery mildew of grapevines. In Grape pest management, 3rd ed. Oakland: University of California, Division of Agriculture and Natural Resources, Publication 3343.

Gubler, W. D., and G. M. Leavitt. 1992. Eutypa dieback of grapevines. In Grape pest management, 3rd ed. Oakland: University of California, Division of Agriculture and Natural Resources, Publication 3343.

Gubler, W. D., J. J. Marois, A. M. Bledsoe, and L. J. Bettiga. 1987. Control of Botrytis bunch rot of grape with canopy management. Plant Disease 71:599–601.

Gubler, W. D., M. R. Rademacher, S. J. Vasquez, and C. S. Thomas. 1999. Control of powdery mildew using the UC Davis powdery mildew risk index. APSnet. www.apsnet.org/online/feature/pmildew.

Gubler, W. D., P. E. Rolshausen, F. P. Trouillas, J. R. Urbez, T. Voegel, G. M. Leavitt, and E. A. Weber. 2005a. Grapevine canker diseases in California. Practical Winery and Vineyard, Feb.-March, Pp. 6–26.

———. 2005b. Research update: Grapevine trunk diseases in California. Practical Winery and Vineyard, Jan.-Feb. www.practicalwinery.com/janfeb05/janfeb05p6.htm.

Gubler, W. D., T. S. Thind, A. J. Feliciano, and A. Eskalen. 2004. Pathogenicity of Phaeoacremonium aleophilum and Phaeomoniella chlamydospora on grape berries in California. Phytopathologia Mediterranea 43:70–74.

Hanna, R., F. G. Zalom, L. T. Wilson, and G. M. Leavitt. 1997. Sulfur can suppress mite predators in vineyards. California Agriculture 51:19–21.

Hopkins, D. L. 1989. Xylella fastidiosa: Xylem-limited bacterial pathogen of plants. Annual Review of Phytopathology 27:271–290.

Khan, A., C. Whiting, S. Rooney, and W. D. Gubler. 2000. Pathogenicity of three species of Phaeoacremonium spp. on grapevine in California. Phytopathologia Mediterranea 39:92–99.

Marois, J. J., A. M. Bledsoe, and L. J. Bettiga. 1992. Botrytis bunch rot. In Grape pest management, 2nd ed., D. L. Flaherty et al., eds. Oakland: University of California, Division of Agriculture and Natural Resources, Publication 3343.

Moller, W. J., and A. N. Kasimatis. 1978. Dieback of grapevines caused by Eutypa armeniacae. Plant Disease Rep. 62:254–258.

———. 1980. Protection of grapevine pruning wounds from Eutypa dieback. Plant Disease 64:278–280.

Munkvold, G. P., and J. J. Marois. 1993. Efficacy of natural epiphytes and colonizers of grapevine pruning wounds for biological control of Eutypa dieback. Phytopathology 83:624–629.

———. 1994. Eutypa dieback of sweet cherry and occurrence of Eutypa lata perithecia in the Central Valley of California. Plant Disease 78:200–207.

———. 1995. Factors associated with variation in susceptibility of grapevine pruning wounds to infection by Eutypa lata. Phytopathology 85:249–256.

Pearson, R. C., and A. C. Goheen. 1990. Compendium of grape diseases. St. Paul, Minnesota: APS Press.

Purcell, A. H. 2005. Xylella fastidiosa web page, UC ANR, www.cnr.berkeley.edu/xylella.

Rea, C., and W. D. Gubler. 2002. The effect of relative humidity on lesion expansion, sporulation, and germination efficiency of Uncinula necator. Proceedings of the 4th International Workshop on Powdery and Downy Mildew in Grapevine, D. M. Gadoury et al., eds. University of California, Department of Plant Pathology, September, 2002.

Redak, R. A., A. H. Purcell, J. R. S. Lopes, M. J. Blua, R. F. Mizell, and P. C. Andersen. 2004. The biology of xylem-fluid feeding vectors of Xylella fastidiosa and their relation to disease epidemiology. Annual Review of Entomomology 49:243–270.

Rolshausen, P. E., and W. D. Gubler. 2005. Use of boron for the control of Eutypa dieback of grapevines. Plant Disease 89:734–738.

Rolshausen, P. E., F. Trouillas, and W. D. Gubler. 2004. Identification of Eutypa lata by PCR-RFLP. Plant Disease 88:925–929.

Rooney, S. N., A. Eskalen, and W. D. Gubler. 2002. Recovery of Phaeomoniella chlamydospora and Phaeoacremonium inflatipes from soil and grapevine tissues. Phytopathologia Mediterranea 40:S351–356.

Rooney-Latham, S., A. Eskalen, and W. D. Gubler. 2005a. Teleomorph formation of Phaeoacremonium aleophilum, cause of esca and grapevine decline in California. Plant Disease 89:177–185.

———. 2005b. Ascospore release of Togninia minima, cause of esca and grapevine decline in California. Plant Health Management, Plant Health Progress, the American Phytopathological Society, Electronic Journal.

Rooney-Latham, S. A., C. N. Janousek, A. Eskalen, and W. D. Gubler. 2009. First report of Aspergillus carbonarius causing sour rot of table grapes (Vitis vinifera) in California. Plant Disease 92:951.

Rumbolz, J., and W. D. Gubler. 2005. Susceptibility of grapevine buds to infection by powdery mildew Ersiphe necator. Plant Pathology 54:535–548.

Sall, M. A. 1980. Epidemiology of grape powdery mildew: A model. Phythopathology 70:338–342.

Sall, M. A., and J. Wyrinski. 1982. Perennation of powdery mildew in buds of grapevines. Plant Disease 66:678–679.

Scheck, H. S., S. J. Vasquez, D. Fogel, and W. D. Gubler. 1998. Grape growers report losses to black foot and young vine decline. California Agriculture 52:18–23.

Scheuerelle, S., and W. Mahaffee. 2002. Compost tea: Principles and practice for plant disease control. Compost Science and Utilization 10:313–338.

Stapleton, J. J., W. W. Barnett, J. J. Marois, and W. D. Gubler. 1990. Leaf removal for pest management in wine grapes. California Agriculture 44:15–17.

Stapleton, J. J., and R. S. Grant. 1992. Leaf removal for non-chemical control of the summer bunch rot complex of wine grapes in the San-Joaquin Valley. Plant Disease 76:205–208.

Thomas, C. S., R. B. Boulton, M. W. Silacci, and W. D. Gubler. 1993. The effect of elemental sulfur, yeast strain, and fermentation medium on hydrogen sulfide production during fermentation. American Journal of Enology and Viticulture 44(2):211–216.

Trouillas, F., and W. D. Gubler. 2004. Identification and characterization of Eutypa leptoplaca, a new pathogen of grapevine in Northern California. Mycological Research 108:1195–1204.

United States Department of Agriculture (USDA). 2004. National Organic Standards Board, Compost Tea Task Force Report. www.ams.usda.gov/nosb/meetings/CompostTeaTaskForceFinalReport.pdf.

University of California, Statewide IPM Program. 2006. Grape pest management guidelines. www.ipm.ucdavis.edu/PMG/selectnewpest.grapes.html.

Úrbez-Torres, J. R., and W. D. Gubler. 2009. Pathogenicity of Botryosphaeriaceae species isolated from grapevine cankers in California. Plant Disease 93:584–592.

———. 2008a. Pathogenicity and epidemiology of Botryosphaeriaceae from grapevines in California. Phytopathologia Mediterranea 48:176.

———. 2008b. Double pruning, a potential method to control Bot canker disease of grapes, and susceptibility of grapevine pruning wounds to infection by Botryosphaeriaceae. Phytopathologia Mediterranea 48:185.

Úrbez-Torres, J. R., G. M. Leavitt, J. C. Guerrero, J. Guevara Lugo, and W. D. Gubler. 2008. Identification and pathogenicity of Lasiodiplodia theobromae and Diplodia seriata, the causal agents of bot canker disease of grapevines in Mexico. Plant Disease 92:519–529.

Úrbez-Torres, J. R., G. M. Leavitt, T. Voegel, and W. D. Gubler. 2006. Identification and distribution of Botryosphaeria species associated with grapevine cankers in California. Plant Disease 90:1490–1503.

Úrbez-Torres, J. R., J. Luque, and W. D. Gubler 2007. First report of Botryosphaeria iberica and B. viticola associated with grapevine decline in California. Plant Disease 91:772.

Vail M. E., and J. J. Marois. 1991. Grape cluster architecture and the susceptibility of berries to Botrytis cinerea. Phytopathology 81:188–191.

Vail, M. E., J. A. Wolpert, W. D. Gubler, and M. R. Rademacher. 1998. Effect of cluster tightness on Botrytis bunch rot in six Chardonnay clones. Plant Disease 82:107–109.

Varela, L. G, R. J. Smith, and P. A. Phillips. 2001. Pierce's disease. Oakland: University of California, Division of Agriculture and Natural Resources, Publication 21600.

Weber, E. A., F. P. Trouillas, and W. D. Gubler. 2007. Double pruning of grapevines: A cultural practice to reduce infections by Eutypa lata. American Journal of Enology and Viticulture 58:61–66.

Whiting, E. C., A. Khan, and W. D. Gubler. 2001. Effect of temperature and water potential on survival and mycelial growth of Phaeomoniella chlamydospora and Phaeoacremonium spp. Plant Disease 85:195–201.

Ypema, H. L., and W. D. Gubler. 2000. The distribution of early season grapevine shoots infected by Uncinula necator from year to year: A case study in two California vineyards. American Journal of Enology and Viticulture 51:1–6.

Insects and Mites in Organic Vineyard Systems

MICHAEL J. COSTELLO

Insect and mite management presents a particular challenge in organic grape production. Because far fewer chemical controls are available to organic growers than to conventional growers, more attention must be paid to preventive tactics such as cultural and biological controls. In addition, because most organically approved chemical controls are not selective—that is, they kill beneficial organisms as well as pests—you must pay particular attention to treatment timing so as to cause the least possible disruption to natural enemies. Also, it is important that organic growers employ the principles of integrated pest management (IPM) in order to keep their pesticide use to a minimum. IPM is a pest management strategy that can be applied to organic, sustainable, or conventional systems and involves sampling, an awareness of economic injury levels, the use of cultural and biological controls as preventives, and the judicious use of chemicals.

There are two ways for a grower to be organic and *not* adhere to IPM principles. The first would be to practice no arthropod pest management at all and hope that the pests take care of themselves naturally. Indeed, there are vineyards in certain pockets of California that for whatever reasons (and these, by the way, are not related to organic vs. conventional management) have very little arthropod pest pressure and are rarely if ever sprayed with insecticides or miticides. However, the vast majority of vineyards do not fit this category. The other way to manage a vineyard organically without following IPM principles is to spray organically approved insecticides preventively and (to compensate for their lower efficacy or residual control) frequently. Neither the "do-nothing" system nor the "excessive-spray" system adheres to the true spirit of organic production, and neither approach is recommended here. The best and most sustainable long-term approach is to incorporate IPM into the organic program.

A more progressive IPM program would incorporate simultaneous management strategies for all pests, recognizing that, often, actions taken against one pest may influence other, unrelated pests. For example, in regions where Pacific spider mite is present, the heavy use of sulfur dust for powdery mildew control leads to a higher density of mites. Also, certain weed species, if present, may serve as alternate hosts for omnivorous leafroller or orange tortrix. An even more holistic IPM program—and one that is seldom undertaken—involves the integration of pest management strategies with cultural practices such as irrigation, fertilization, and canopy management. For example, over-irrigated vines provide lush vegetation that leafhoppers like, whereas severely stressed vines are more prone to Pacific spider mite. Specific integrated tactics will be discussed in depth in the text that follows. For further information on IPM principles and applications, see the sources given in the References at the end of this chapter.

INTEGRATED PEST MANAGEMENT

There are five components to IPM for arthropods that any grower needs to address in some fashion:

- identification of the pest
- sampling (monitoring)
- economic injury level (EIL)
- cultural and biological controls as preventive measures
- chemical controls

Identification. Correct identification of the insect or mite is critical if you are going to make the right treatment decisions. There are many multi-legged creatures in the vineyard; some may cause injury, but most are either innocuous or beneficial to the crop. A useful tool is a good quality 10× hand lens. Some may be tempted to use a 15× or stronger lens, but this is really not necessary and may even be undesirable. Typically, the higher a lens's magnification, the smaller its field of vision. A pest control adviser (PCA) might want to invest in a 40× microscope to keep in the office or laboratory. If a grower or PCA is unsure of the pest's identification, it is a good idea to contact a UC Farm Advisor at the local county's Cooperative Extension office or to send a specimen to the county Department of Agriculture office for identification.

Sampling (monitoring). It is important for a grower or PCA to know what types of arthropods are present in the vineyard and at what densities, and the only way to find this out is to monitor (take samples) and keep records of the results. Sampling should be undertaken as frequently as is feasible. Weekly sampling usually works the best in terms of balancing the need for good information with the cost and effort of taking samples. Almost all samples for arthropods and for diseases (see chapter 8) can be taken during a single weekly visit. The only situation that might require more frequent sampling would be an outbreak of Pacific spider mites.

The number of samples you have to take in order to get a precise estimate of the pest population density depends on how variable the counts are in the samples. "Precision" in this sense refers to the degree of variation (or error) among the samples. In theory, for example, if there were no variability in the samples taken and the counts of leafhopper nymphs on every leaf sampled were always the same, then a one-leaf sample is all you would need to take for a precise sample. In reality, however, there is variability, and you have to compensate for that by taking more samples. Then again, the cost of taking the samples is always a concern. The key to good monitoring for pest management decision making is to balance the precision of the sampling estimate with the cost of taking the samples.

Here is a quick and easy way to estimate the proper number of samples to take. The formulas involved are simple and can be programmed into a spreadsheet on a desktop or laptop PC or a handheld PDA (personal digital assistant) for leafhoppers, mites, and any other arthropods for which you want to record a count. The first step is to use a quick, effective method to estimate your sampling precision, which is a measure of the variation (standard error or SE) among the counts in the sample, relative to the average (mean or $\bar{x}$) count. Most spreadsheet programs have the standard error function. If yours does not, use this formula:

$$SE = \text{standard deviation} \div \sqrt{n}$$

where n 5 the number of samples taken.

With these two parameters, we can calculate the precision (D) as follows:

$$D = SE \div \bar{x} * 100$$

A value for D of less than 25 percent is considered acceptable. What this means in practical terms is that the estimate of pest density has a 25 percent margin of error (i.e., the mean plus or minus 25%). So, for example, if the average leafhopper nymph count per leaf is 10 and D = 25 percent, then the estimate range is from 7.5 nymphs per leaf (10 − 2.5 = 7.5) to 12.5 (10 + 2.5 = 12.5) nymphs per leaf.

If D is greater than 25 percent, you may need to take more samples just to get a narrower range of precision. However, if you have taken more than 30 samples and D is still above 25 percent, you have to make a decision: If the high figure in the estimate range (12.5 in the example) is below the economic injury level (EIL, see next paragraph), it probably is safe for you to stop sampling for now and re-sample after a few days or a week (a few days for Pacific spider mite, which can reproduce rapidly, but a week for leafhoppers and other slower reproducers). If the high figure in the estimate range is at or above the EIL, the conservative course of action is to treat immediately.

Economic injury level (EIL). In theory, the EIL is an indication of how much injury a plant can sustain before the crop incurs economic damage (i.e., a loss of yield or quality). In reality, though, the EIL refers to pest density. It represents the point at which the pest's population becomes so high that the amount of economic injury caused is equal to the cost of control, and any further injury means a monetary loss for the grower. The injury level at which the grower would need to apply chemical treatment to actually prevent economic damage is somewhat lower than the EIL, since it takes a certain amount of time after monitoring for the grower to prepare the spray machinery, apply the pesticide, and let the pesticide take effect. The EIL is the weakest element in the IPM model, partly because it is dependent on so many variables and partly because the research

10 VINEYARD ARTHROPOD PREDATOR GROUPS YOU SHOULD RECOGNIZE

At the crux of IPM is the need to maximize the preventive pest controls: cultural and biological controls. As such, a grower needs to encourage the presence of beneficial arthropods in the vineyard as much as possible. Although many of the beneficial parasitoids are difficult to identify, most predators are easily recognizable. Here are the top ten predator groups that an astute IPM practitioner should be able to recognize on sight.

Spiders. Overall, spiders make up the most consistent and abundant predator group found in California vineyards. They are very diverse, representing over a dozen families and scores of species. Some of the most important families are

- Miturgidae (long-legged sac spiders). The most common members of this family are the *Cheiracanthium* spp., commonly known as agrarian or yellow sac spiders (Figure 9.1)

- Corinnidae (corinnid sac spiders) (Figure 9.2)

- Salticidae (jumping spiders)

- Oxyopidae (lynx spiders)

- Theridiidae (cobweb weavers) (Figure 9.3)

- Araneidae (orb weavers)

Spiders are beneficials in that they eat and consume a wide variety of prey, but they have their limitations: most species have only one generation per year, and they do not feed solely on pests. However, field observations have shown that spiders do consume vineyard insect pests. These include *Trachelas pacificus* (Corinnidae) and *Cheiracanthium inclusum* (Miturgidae) preying on omnivorous leafroller and *Metaphidippus vitis* (Salticidae) and other jumping spiders preying on grape leafhopper nymphs (but not variegated leafhopper nymphs).

Phytoseiid mites. Phytoseiids are predatory mites and can be recognized by their shiny, oval-shaped body (Figure 9.4). The most common vineyard phytoseiid statewide is *Galendromus occidentalis,* otherwise known as the western predatory mite. It is a specialist, feeding on spider mites, tydeiid mites, and eriophyid mites. On glabrous grape varieties (those having very few leaf hairs) such as Grenache, *Galendromus* can quickly overtake a spider mite population if 50 percent of the leaves sampled have at least one predatory mite. On hirsute (hairy) varieties such as Zinfandel, it takes longer for the predators to reduce spider mite density. In coastal counties of California, a generalist phytoseiid species, *Neoseiulus californicus,* is quite common.

Sixspotted thrips (*Scolothrips sexmaculatus* [Pergande]). Not all thrips are plant eaters; the sixspotted thrips (Figure 9.5) is a specialist predator of spider mites, feeding on all stages, including the eggs. Adult thrips are recognizable by three dark spots on each wing. Immatures are white, with a slightly more bulbous abdomen than western flower thrips nymphs. Sixspotted thrips are voracious feeders and can reduce a spider mite population even more rapidly than *Galendromus*. Apparently they do not overwinter in the vineyard, so they typically are not found until mid- to late summer.

Black hunter thrips (*Leptothrips mali* [Fitch]. This is a very small, jet black thrips (Figure 9.6) that is common in the Sierra Foothills and North Coast regions. It probably feeds on thrips, mites, and early instar leafhoppers.

Minute pirate bugs (*Orius* spp.). These small true bugs with a black ✕ pattern on their wings (Figure 9.7) can be quite common in North Coast vineyards, where they probably feed on thrips, mites, and leafhopper nymphs. The most commonly found stage is the nymph, which is orange and teardrop shaped.

Lacewings (*Chrysopa* and *Chrysoperla* spp.). Although lacewings (Plate 9.8) are well recognized as generalist predators, they are not very abundant in most vineyards. Lacewing eggs are a fairly common sight, but for whatever reason, the population does not seem to build to very high numbers. Lacewing larvae are very fond of mealybugs, although high lacewing densities have not been associated with infestations of grape, obscure, longtailed, or vine mealybugs. Lacewing larvae have occasionally been observed feeding on leafhopper nymphs.

Spider mite destroyer (*Stethorus picipes* Casey). This is a very small, black, hirsute lady beetle (family Coccinellidae) (Figure 9.9) that specializes on spider mites. It is most common in North Coast vineyards.

Damsel bug (*Nabis* spp.). Also known as nabid bugs, these true bugs (Figure 9.10) are probably most commonly found in spring, though their densities throughout the state are very low in vineyards. Little has been observed of their feeding habits in the vineyard, but in other cropping systems they eat lepidopteran larvae and other soft-bodied arthropods.

Convergent lady beetle (*Hippodamia convergens* Guérin-Méneville). This California native (Figure 9.11) is the most commonly found lady beetle in California vineyards, recognizable by the two convergent white stripes on its thorax. It can be a very abundant vineyard predator in the spring, especially if a cover crop is present. However, no observations have yet been made as to which specific vineyard arthropod pests it feeds upon.

Big-eyed bug (*Geocoris* spp.). A true bug (Figure 9.12), it feeds on thrips, whiteflies, and spider mites. Big-eyed bug is rarely found in vineyards.

Figure 9.1. Cheiracanthium spider. *Photo* by Michael Costello.

Figure 9.4. Phytoseiid mites. *Photo* by Michael Costello.

Figure 9.2. Trachelas spider. *Photo* by Michael Costello.

Figure 9.5. Sixspotted thrips (*Scolothrips sexmaculatus* [Pergande]). *Photo* by Jack Kelly Clark.

Figure 9.3. Cobweb weaver spider (*Theridion* sp.). *Photo* by Michael Costello.

Figure 9.6. Black hunter thrips (*Leptothrips mali* [Fitch]. *Photo* by Jack Kelly Clark.

Figure 9.7. Minute pirate bug (*Orius* spp.) adult and nymph. *Photos* by Jack Kelly Clark and Michael Costello.

Figure 9.10. Damsel bug (*Nabis* spp.). *Photo* by Jack Kelly Clark.

Figure 9.8. Lacewing (*Chrysopa* or *Chrysoperla* spp.). *Photo* by Jack Kelly Clark.

Figure 9.11. Convergent lady beetle (*Hippodamia convergens* Guérin-Méneville). *Photo* by Jack Kelly Clark.

Figure 9.9. Spider mite destroyer (*Stethorus picipes* Casey). *Photo* by Jack Kelly Clark.

Figure 9.12. Big-eyed bug (*Geocoris* spp.). *Photo* by Jack Kelly Clark.

community devotes only a very small amount of money and time to its study. A given pest's EIL can fluctuate over time, even within a season, and is dependent on the value of the crop, the cost of the control measure, the crop variety's susceptibility to the pest (this can be influenced by environmental conditions), and the efficacy of the control measure. None of these factors is static. Because its derivation is so complex, in most real-world cases the EIL is effectively a matter of the grower's or PCA's subjective level of tolerance, based on the individual's personal experience with that pest in a given vineyard. Keep in mind that when we refer to EILs in this chapter, they are only estimates and must be adjusted to suit any specific vineyard and, frankly, each practitioner's tolerance level.

Cultural and biological controls as preventives. "Cultural controls" are manipulations of the environment that help to lessen a pest's population density or mitigate the damage it can do. For example, an important cultural control for lepidopteran pests in many cropping systems is the destruction of harvested crop residues in fall to reduce overwintering habitat. In the vineyard, this is done by burying or shredding unharvested grape clusters, which are potential habitat for overwintering omnivorous leafroller and orange tortrix. Cultural controls also include practices that help the plant defend itself against insects and mites as well as pathogens. This includes soil fertility and water management. Some cultural controls can be part of a long-term strategy against a variety of arthropod pests, and some may cross over into the category of biological control: for example, planting hedgerows of plants that provide an alternate food source for natural enemies. These may be thought of as "living insectaries" because many parasitic Hymenoptera and Diptera will live longer and lay more eggs if they have a plant nectar source nearby. The same goes for predators such as lacewings, if you provide a source of pollen. Parasitic Hymenoptera that attack leafhoppers and mealybugs on grapes are less likely to take advantage of these floral resources since they can feed on honeydew produced by their hosts. Ideas for nectar- and pollen-producing plants that are appropriate for a vineyard insectary, border planting, or hedgerow can be found in Thrupp et al. (2008).

"Biological controls" are pathogens, predators, and parasitoids that attack the pest and reduce its population density. They are also referred to collectively as "natural enemies" or, simply, "beneficials." There are natural enemies present in California that have the potential to control nearly all of the state's grape arthropod pests. Natural enemies can also be purchased and released, but this is a more expensive alternative than most available pesticides, even those that are organically approved. The effectiveness of releases is often difficult to assess, and releases need to be preventive and made well in advance of pest population buildup because their effects are not immediate.

Chemical controls. Chemicals available to organic grape growers are limited. Organic growers need to make sure that the pesticides they use are approved by the Organic Materials Review Institute (OMRI, online at http://omri.org) and that they are registered for use on grapes by California's Department of Pesticide Regulation (DPR, online at http://www.cdpr.ca.gov). The use of chemical controls should only be undertaken when the insect or mite population nears the EIL. If, after all preventive measures have been taken, the pest population continues to increase, then chemical control is the only remaining way to reduce it. The difficulty for organic growers is that most organically approved chemical controls are not selective, and therefore natural enemies and nontarget arthropods may suffer more harm than they would from selective but synthetic materials. Organically acceptable chemical controls must be applied with extreme caution.

In general, the chemical tools available to the organic vineyard manager are insecticidal soaps, oils, botanicals, minerals, and microbials. Insecticidal soaps are potassium salts of fatty acids that react with the waxy cuticle of soft-bodied arthropods (the cuticle protects arthropods from water loss). Soaps also reduce the surface tension of water. The result is either the drowning of the arthropod, if water enters into its spiracles (air openings), or desiccation, once the waxy cuticle breaks down. There is just one formulation of insecticidal soap registered for use on grapes. Although you may be tempted to use a household soap, this is not recommended: not only is it illegal (since household soaps are not registered for use on agricultural crops), but the impurities in household soap may be phytotoxic to the vine.

Mineral oils are highly refined to remove impurities, and go by names such as "narrow-range," "horticultural," and "superior." Botanical oils include the common vegetable oils (e.g., soybean, canola, and cottonseed oils) and the essential oils (which contain traces of phytochemicals such as terpenes) (e.g., rosemary, clove, and spearmint). Botanical oils are very pure and there is little chance of phytotoxicity. Still, for both mineral and botanical oils, you should take care not to apply them when the temperature is too high (check the label for specific temperatures). The effect of a mineral or vegetable oil is to cover the spiracles and suffocate the arthropod. The essential oils will do this as well, and the phytochemicals they contain may also be toxic (which is why their label rates are generally lower than those of other botanical oils). Particular care must be taken with oils to avoid applying them too soon before or after a sulfur application, as the interaction between the two can lead to phytotoxicity. The label will have a recommended safe interval period.

The botanicals available for organic use are pyrethrum (pyrethrin) and neem, and these have different modes of action. Pyrethrum is a nerve (axonic) poison, and neem blocks the action of ecdysone, an insect hormone that is essential for molting. They are broad-spectrum materials, meaning that along with their negative impact on pests they may also have a negative effect on beneficials.

Minerals for insect and spider mite management include lime sulfur and kaolin. Lime sulfur (calcium polysulfide) is effective against spider mites and mealybug crawlers, but because it is caustic it can only be used in the dormant season. Kaolin is a type of clay and is applied as a fine dust. It can be used as a repellent for leafhoppers and sharpshooters.

The most widely used microbial is Bt, a toxin derived from the bacterium *Bacillus thuringiensis,* which is effective on immature Lepidoptera. When a caterpillar ingests Bt, it destroys specialized cells (epithelial cells) that line the insect gut, leading to severe digestive problems. The only other microbial registered for organic use on grapes in California is spinosad, a nerve poison also used on caterpillars as well as thrips.

A major issue with these organically approved materials, especially the soaps and oils, which are contact materials, and Bt, which must be ingested, is the amount of water used to apply the material. A reasonable amount to apply to a vineyard from budbreak through the first several weeks is 50 to 60 gallons per acre (4.6 to 5.6 hectoliters/ha). Once shoot length has increased beyond 18 inches (46 cm) and the canopy has filled out, you may have to increase the spray volume to 100 to 150 gallons per acre (3.8 to 5.7 hectoliters/ha) to ensure adequate coverage, especially if contact materials are being used. In order to achieve this, you can use higher-volume nozzles and reduce your tractor speed to 2 mph (3.2 km/hr) or less.

MANAGEMENT GUIDELINES

What follows is a set of management guidelines for the most important arthropod pest groups affecting grapes in California. Although there are perhaps a dozen additional minor grape pests, they only occasionally reach numbers high enough to warrant control measures. Minor pests include thrips, lecanium or brown apricot scale, grape whitefly, cutworms, grape bud beetle, and eriophyid mites. More information on these and other pests is available from the UC Statewide IPM Program (online at http://www.ipm.ucdavis.edu).

Spider Mites (Tetranychidae)

"Spider mite" is the term applied to mite species in the family Tetranychidae, all of which feed on plants and produce some degree of webbing. Spider mites are not too small to be seen with the naked eye, but it is best to monitor them with a hand lens. Only two spider mite species cause consistent and significant injury to grapes in California: Pacific spider mite (PSM) (*Tetranychus pacificus* McGregor) (Figure 9.13) and Willamette spider mite (WSM) (*Eotetranychus willamettei* [McGregor]) (Figure 9.14). The twospotted spider mite (*Tetranychus urticae* Koch) is occasionally found on grapes in California and even more rarely does it reach pest densities.

Description. In general, PSM is larger, hairier, and more colorful than WSM and produces considerably more webbing. Be cautioned that the mite's color alone is not an adequate identifier but should be weighed in combination with other characteristics such as the type of injury it causes and its distribution on the leaf and vine. PSM is typically yellow-orange but can also be pale yellow or green. WSM tends to be pale white to yellow and is occasionally pale green. The adult female PSM is about 0.5 mm long, slightly larger than WSM. Adult males of both species are about half the size of females and can be recognized as males by their pointed abdomen. Absolute certain identification can only be made by examining the male reproductive organ (the aedeagus) under high-powered (60×) magnification.

Figure 9.13. Pacific spider mite (adult female). *Photo* by Michael Costello.

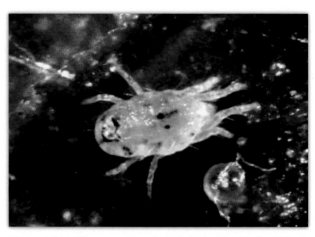

Figure 9.14. Willamette spider mite (adult female). *Photo* by Michael Costello.

Both PSM and WSM lay smooth, spherical eggs about 0.2 mm in diameter. At first the eggs are transparent, but they turn opaque as the embryo develops within. A newly hatched mite, called a larva, is not much larger than the egg; you can identify it because it has six legs (rather than the eight legs of older instars) and red eyes (Figure 9.15). The next two instars (developmental stages) are called nymphs. Nymphs have eight legs and begin to show visible food spots, which are actually accumulated waste products that are stored in lateral sacs in the body. Food spots tend to occur along the sides of the mite and are larger and more concentrated in PSM than in WSM. PSM nymphs also have two food spots at the end of the abdomen.

Injury and damage. Spider mites feed on individual leaf cells, puncturing them and removing the contents. Initial distribution for WSM on a plant tends to be along the primary and secondary veins of leaves, so injury follows this pattern, appearing as yellowing on white varieties such as Chardonnay (Figure 9.16) and reddening on red varieties such as Merlot (Figure 9.17). As WSM population density increases, discoloration increases, but leaves do not burn. PSM populations tend to cluster more on the leaf, preferring folds and crevices (Figure 9.18). Initial injury from PSM, then, is more concentrated, with yellowed leaves on both red and white varieties and, eventually, reddened leaves on red varieties (Figure 9.19). Under high PSM population pressure, leaves will turn necrotic. If injury to photosynthetically active leaf areas is severe enough, there can be a delay in berry sugar accumulation, a loss of

vigor, or a decrease in yield. A substantial decrease in vigor can even reduce the following year's yield. In some coastal areas, WSM will emerge from its overwintering habitat after budbreak and cluster on young shoots in high densities, a condition known as "spring mites." While this looks alarming, there is no evidence that the phenomenon causes any long-term damage. It is likely that as the vines grow, the mites simply disperse over the increasing vine area, diluting any potential negative effects.

Distribution on the vine. Both WSM and PSM overwinter as adults under the vine's bark. For this reason, their initial infestation in spring is on the basal leaves. WSM populations migrate rather slowly over the course of the season from the lower leaves to the midcane area, and they are never found in high numbers on the shoot tips. PSM will remain on the lower and midcane leaves until the weather warms up. Then, as daily temperature highs approach and top 95°F (35°C), PSM migrate rapidly to the upper portions of the canopy, where they cluster on the shoot tips.

Regional distribution. WSM is present throughout California and is the primary pest mite for most of the North Coast, parts of the Central Coast (north of King City and south of Paso Robles), the Sierra Foothills, the western half of the San Joaquin Valley, and parts of eastern Tulare and Kern Counties. Wherever PSM is present it tends to become the dominant pest. It is found on the eastern half of the San Joaquin Valley, in the Sacramento Delta, and in the warmer regions of the North Coast (Lake County and the upper regions of Sonoma and Napa Counties) and the

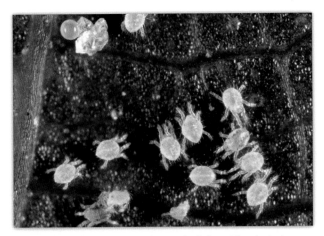

Figure 9.15. Spider mite larvae have only six legs. *Photo* by Jack Kelly Clark.

Figure 9.16. Willamette spider mite injury on Chardonnay. *Photo* by Michael Costello.

Figure 9.17. Willamette spider mite injury on Merlot. *Photo* by Michael Costello.

Figure 9.18. Pacific spider mite injury on Chardonnay. *Photo* by Michael Costello.

warmer regions of the Central Coast (south of King City through Paso Robles).

Development. WSM has a moderate and fairly regular population growth rate that does not increase with extremely hot weather. It often appears shortly after budbreak in March but rarely reaches populations high enough to cause damage until July or later. PSM usually first appears later in the season (May or June), but its population growth rate is moderate until hot weather (95°F [35°C]) sets in, at which point its rate of expansion accelerates rapidly.

Sampling and EIL. Sampling for mites represents a particular challenge because of their small size and high population density. As with all sampling, the challenge is to strike a balance between obtaining enough information to make a good decision and keeping collection costs to a reasonable level.

On east-west rows, PSM should be sampled from the south side of the vine; on north-south rows, you

Figure 9.19. Pacific spider mite injury on basal leaves of Merlot. *Photo* by Michael Costello.

can sample both PSM and WSM from either side. Early in the season (from shortly after budbreak to bloom), sample for mites by looking on the first few basal leaves. As the season progresses, populations will migrate to newer leaves, so you will need to sample from farther up the cane. By midseason, samples should be taken from the lower- to midcane area, about the fourth to eighth nodes. The most obvious thing to look for is leaf discoloration, which is detectable from the top side of the leaf. The mites will be on the underside of the leaves. They can usually be seen with the naked eye, but it is best to confirm with a 10× hand lens. Eggs usually cannot be seen without magnification. Take samples once a week, and when density of PSM is nearing the EIL increase your sampling frequency to every 3 or 4 days. In areas where leafhoppers (see below) are also a concern, you can sample and record both pests together.

The most common method of sampling for spider mites on grapes is the unregimented method, and decisions about the number of samples to take and when to apply control measures are largely based on personal experience. This method is more art than science. Its main advantage is that it is very quick. The downside, though, is that it requires an experienced eye, and often the actual decision to treat is not based on a hard estimate of the population. A PCA will usually work with a given block that is uniform in variety and age and take a number of leaf samples (sometimes a few, sometimes many), looking for leaf injury and mite colonies. Often, samples are taken from areas of the block where population increases have begun in the past—for example, the weak spot where PSM populations have tended to build up. Some PCAs make a mental count of per-leaf mite density and can recognize at a glance whether a leaf has, say, 5, 10 or 20 mites, but most PCAs simply record the number of leaves that have colonies at all. Predatory mites are noted when seen, but little time is generally spent searching leaves for them. Sometimes a PCA will delay a treatment decision if predatory mite density is relatively high. The decision to spray is based either on the degree of leaf discoloration, the number of leaves with colonies, or a density estimate. The decision is also based sometimes on weather or time of the season (e.g., PSM tends to increase rapidly after July 1, or with increasing temperatures).

A presence-absence sampling method for PSM (but not for WSM) outlined in *Grape Pest Management* (Flaherty et al. 1992) and on the UC IPM website (http://www.ipm.ucdavis.edu) recommends breaking each block into quadrants and taking one leaf from each of five randomly selected vines per quadrant, for a total of 20 leaves. Each leaf is scanned with a hand lens, and the presence of PSM and predatory mites is recorded on a scoresheet. The treatment decision is based on the percentage of leaves that score positive for PSM and the number of leaves with predatory mites. For example, if 50 percent of the leaves have PSM and 1 of the 20 leaves sampled has a predator, a decision to spray could be delayed until the next sampling. However, if the PSM infestation were 50 percent and no predators were found, the recommendation would be to spray immediately.

The advantage of this method is that it ensures that the field is sampled uniformly and randomly, which is necessary if the field's variability and its history of spider mite infestations are not known. Also, the presence-absence method takes less time than an actual mite count. The downside to this method is that it is quite rigid, and if a PCA is familiar with the particular field, partitioning it into quadrants is not necessary because he or she can concentrate initial sampling efforts in sections where spider mite outbreaks have initiated in the past. Also, it is more important for you to get a reliable estimate than to collect a fixed number of samples. If conditions are fairly uniform, fewer than 20 samples may suffice; if there is a lot of variability, you may need to take as many as 30 samples. Finally, the presence-absence method does not allow analysis of the cumulative effects of mite feeding.

A third sampling method is to count the number of mites per leaf. Counting is the only way to evaluate the cumulative effect of spider mite infestations. The advantages of this method and cumulative mite-day analysis include an increased accuracy for the population density estimate. Also, with this method the treatment decision does not rely on observations at a single point in time (a static analysis) but instead integrates mite densities over time (i.e., it analyzes the cumulative effect). The mite brushing machine used in this approach also facilitates predatory mite identification and density estimation. The disadvantages are the expense of the brushing equipment and of the additional time it takes to brush the leaves and count the mites.

In most cases, it is too tedious and time consuming to count every mite with a hand lens. An experienced sampler can, however, make a quick and crude estimate of the mites on any given leaf. Mite counting can also be done with a leaf brushing machine made by Leedom Enterprises of Mi-Wok Village, California (Figure 9.20). The machine can be taken into the field and plugged into the 12-volt power outlet of an automobile, using a power inverter. Typically, 10 leaves are brushed onto one plate, and the mites on 10 or 20 percent of the plate are then counted under a binocular microscope. That count is extrapolated into an estimated number of mites per leaf (Figure 9.21). The grower or PCA then estimates the cumulative effect of mite feeding by calculating "mite-days." A mite-day is the equivalent of one mite per leaf for one day. So 10 mites per leaf for one day would be 10 mite-days, and 10 mites per leaf for seven days would be 70 mite-days. To get cumulative mite-days, you have to average the counts of two successive samplings and then multiply that mean by the number of days from the first to the second sampling. A typical mite-day table might look something like Table 9.1.

Research is still being conducted on EILs using cumulative mite-days, and it appears likely that they differ by grape variety. With a standard ("sprawl") trellis and minimal leaf removal, the EIL for WSM on Zinfandel (considered a fairly susceptible variety) is probably somewhere between 1,000 and 2,000 mite-days, whereas on Chardonnay it is higher, perhaps 1,500 to 2,500 mite-days. On Thompson Seedless, considered a somewhat tolerant variety, the EIL for PSM is probably between 2,000 and 3,000 mite-days.

Control. Because they are very sensitive to the condition of the plant and changes in the environment, spider mites can often be managed effectively with cultural and biological controls. This course takes determination and no small amount of patience, however, since it may take a grower several seasons to achieve the proper balance between vine condition and biological controls.

Cultural controls. Mites do better under dusty conditions (why this is so is not known), and it is very common for vines that receive a lot of dust to have more mites. Typically, the vines with the most dust are the vines that border roadways. Simply minimizing dust from unpaved roadways can be an important element in spider mite management.

There are a number of ways to reduce roadway dust, and they vary significantly in cost. Paving is the most effective option, but by far the most expensive. Gravel is also very effective, but again it is very expensive. Oil is effective and costs less, but it has a short effective life span of only a season or two. Lignins (or lignin sulfonates) are longer lasting—say, 10 years—and are cost effective if amortized over this period. The least expensive option, water, is also the shortest lasting. It is not uncommon for a ranch to hire a worker just to drive a water tanker around all day and keep the roads wet, so long as the ranch is sufficiently large (Figure 9.22).

PSM appears to respond to severe plant water stress with rapid development and increased reproduction. If WSM has a similar response to water stress, it has

Figure 9.20. Leaf brushing machine. *Photos by Michael J. Costello.*

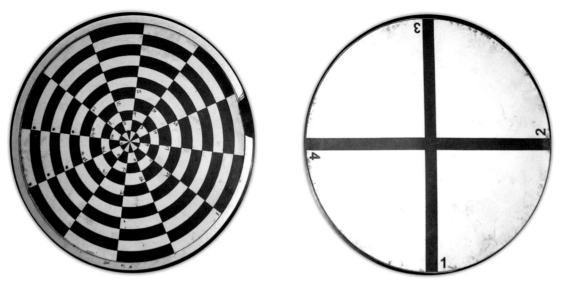

Figure 9.21. Two types of counting grid for estimating mite populations: The wedge and the cross. *Photo* by Michael J. Costello.

Table 9.1. Estimating mite-days

Week	Mites per leaf	Average mite-days per day (week 1 + week 2) ÷ 2	Average mite-days per week (average mite-days per day x7)	Cumulative mite-days
Week 1	0	–	–	–
Week 2	4	2	14	14
Week 3	6	5	35	49
Week 4	10	8	56	105
Week 5	16	13	91	196

not been obvious enough to be noted. Because there has been no formal study of the effect of water stress on PSM or WSM, the exact plant water status most favored by the mites is not known. It is probably safe to say, though, that highly stressed vines (vines with over 18 bars [–1.8 mPA] of leaf water potential) are more likely to trigger increased PSM activity. If you practice deficit irrigation in a vineyard where PSM is a concern, then, you may want to keep leaf water potential to 14 to 16 bars (–1.4 to –1.6 mPA) or less, just to stay on the safe side.

Cover cropping can be helpful or harmful in terms of managing spider mite density, depending on how the cover crop is managed. If the cover crop competes with the vines for water, it could lead to an increase in water stress that might favor PSM. On the other hand, a well-managed cover crop can improve soil structure and improve water penetration, thereby improving vine water status. Some types of cover crop can support alternate

prey for predatory mites (e.g., other mite species), but you cannot rely upon this. For example, some cover crops—especially legumes such as clovers and vetches—can be colonized by twospotted spider mite, and a buildup of predatory mite populations in the cover crop often follows. The presence of twospotted spider mite cannot be guaranteed, though, and in some years populations will be low or lacking altogether. Even if twospotted spider mite is present at a high density on a cover crop, there is no research to indicate with any certainty that its presence will lead to better biological control of either PSM or WSM on the vines.

Formal studies on the effects of vine nutrient concentration or nutritional stress on PSM or WSM are rare, but evidence and observations suggest a link between vine nutrition and spider mite density. Typically, vineyards with a low overall nutrient concentration have a lower density of spider mites. This may be related to different kinds of nutrients,

Figure 9.22. Watering roadsides reduces dust and aids in spider mite management. *Photo* by Michael Costello.

but it is very likely that nitrogen plays a role since other spider mite species, including twospotted spider mite, respond positively to high leaf nitrogen concentrations. It may therefore be wise for growers to be moderate in their use of nitrogen fertilization, organic and otherwise. Consider keeping bloomtime petiole concentration of NO_3 between 500 and 700 ppm (less than 1% total N).

PSM does well under conditions of frequent sulfur dust application, but no similar connection has yet been shown experimentally for WSM. In regions where PSM is found, the use of alternatives to sulfur dust may be beneficial. This, of course, is not good news for organic growers, who rely on sulfur as one of the few options for effective powdery mildew management. However, micronized (wettable) sulfur does not seem to have the same effect on mite populations, perhaps because it is applied at lower rates (i.e., 3 to 4 lb/ac [3.4 to 4.5 kg/ha] versus 10 to 15 lb/ac [11.2 to 16.8 kg/ha] for sulfur dust). Timing may be important, too: there is some evidence that early season (i.e., pre-bloom) sulfur dusting leads to higher PSM densities than mid- or late-season dusting. Additional alternative treatments for mildew control that do not exacerbate PSM are bicarbonates and oil, but good coverage is essential to make these treatments effective (see chapter 8).

Biological controls. The most important thing you can do to encourage the biological control of spider mites in a vineyard is to avoid using chemicals that discourage natural enemies. Like synthetic chemicals, some organically approved controls such as soaps and oils have a negative affect on beneficials

as well as injurious mites. If you minimize chemical use, your vineyard may even develop an indigenous population of predatory mites. The western predatory mite (*Galendromus occidentalis* Nesbitt) is widespread throughout California and quite common where broad-spectrum chemical controls are kept to a minimum. Other predatory mite species may be found in specific regions, including *Neoseiulus californicus* (McGregor) on the coast and *Metaseiulus mcgregori* (Chant) in the San Joaquin Valley. Predatory mites (*G. occidentalis*) can be purchased, typically on flats of bean plants where they have been reared. To release the mites, you place the plants, usually by hand and one at a time or several at once, onto the head of a vine. To be effective, a release must be fairly large.

There are two strategies for release: single release (inundation) in the season and multiple releases (augmentation) over the course of the season. A one-time release will probably require at least 20,000 mites per acre (about 50,000 per hectare) in order to have a chance of success. Augmentative releases can be made at a lower rate many times a season; say, 2,000 per acre (5,000 per hectare) once a week for 10 consecutive weeks. In either case, the expense of the releases will far surpass that of even the most expensive organically approved chemical controls. The other biological control agent that can control either PSM or WSM is sixspotted thrips (*Scolothrips sexmaculatus* [Pergande]). This thrips preys upon mite eggs and mites in all growth stages. When they arrive in a vineyard, they quickly reduce the density of spider mites. Unfortunately, sixspotted thrips do not overwinter in the vineyard, and there is no predicting when they will appear. Sixspotted thrips can also be purchased from commercial insectaries.

Chemical controls. The only organically approved materials that are effective against mites are soap and oils. Soap works by dissolving the mite's waxy cuticle and by lowering the surface tension of water that hits the mites. Oil covers and suffocates them, and phytochemicals in the essential oils can act as direct toxins. The mineral and vegetable oils available to organic growers are typically applied at dilutions of 1 to 2 percent; essential oils are recommended at 0.5 percent. However, the efficacy on WSM and PSM of essential oils applied at this rate has been erratic and needs further study. Soap and oils are contact materials, so coverage is very important. Because 100 percent coverage is almost impossible, several treatments at intervals of 5 to 7 days are necessary to clean up an infestation. Remarkably, water by itself will kill mites; because they have only one pair of spiracles, they are very sensitive to inundation. Water does not, however, kill the mite eggs. Although time consuming and expensive, frequent treatments with water, with good coverage, can control a spider mite population while probably having the lowest negative effect on natural enemies. Sometimes the same effect can be achieved with sprinkler irrigation, although this is probably more effective early in the season when the canopy is light.

Leafhoppers (*Erythroneura* spp.)

Erythroneura is the genus of two insect species that cause injury to grapes in California: the western grape leafhopper (WGLH, *Erythroneura elegantula* Osborn) and the variegated leafhopper (VLH, *Erythroneura variabilis* Beamer). The family name for leafhoppers is Cicadellidae, a group that includes sharpshooters (described later in this chapter), but in grape pest management what we commonly refer to as "leafhoppers" are the *Erythroneura* species. Leafhoppers differ from other well-known hemipteran grape pests such as mealybugs and sharpshooters in that the leafhoppers do not feed on vascular tissue and do not transmit disease. Like most other Hemiptera, leafhoppers do eliminate excess sugars in the form of "honeydew," although they produce a lot less of it than their cousins the mealybugs.

Description. The two leafhoppers, WGLH and VLH, are very similar in size and shape, and are only reliably distinguished from each other by their coloration and by the placement of their eggs. Each species hatches as a small, 1 mm long white nymph. WGLH stays whitish throughout its five instars (Figures 9.23 and 9.24), although it takes on a yellowish hue as it develops. VLH begins to turn yellow-orange late in the first instar (Figures 9.25 and 9.26), dark orange by the second instar, and burnt orange in the fifth instar, by which time it is about 3 mm long. After it molts to adulthood, WGLH shows a yellow-orange pattern against its pale background and can be distinguished by the two black spots on its thorax. VLH adults have a white-orange pattern against a dark background and two prominent white patches on the outer middle fringes of their wings.

WGLH females lay kidney-shaped eggs just under a few layers of leaf tissue. The eggs can be seen on the leaves' lower surface using a microscope or 10× hand lens. VLH females lay cigar-shaped eggs deep within leaf tissue. These can only be seen under a microscope or 10× hand lens when light is transmitted through the leaf.

Injury and damage. Leafhoppers feed on the middle region of the leaf (the mesophyll), piercing individual plant cells with their needle-like stylet. When many cells in an area have been destroyed, small white spots emerge on the leaves in a mottled pattern (Figure 9.27). Injury from both leafhopper species increases over the course of the season; in addition, high late-season VLH population densities cause marginal necrosis, known as "hopper burn."

Figure 9.23. Fifth instar Western grape leafhopper. *Photo* by Michael Costello.

Figure 9.24. Adult western grape leafhopper. *Photo* by Michael Costello.

Leafhopper honeydew falls onto grape berries, forming a medium for sooty molds. For wine and raisin grape growers this is not a concern, but on table grapes this "spotting" is cosmetic damage and will cause the grapes to be downgraded, lowering their market value.

Leafhoppers are aptly named because as adults they make short hop-flights when they are disturbed. This can be a major annoyance to field-workers when leafhoppers fly into their eyes, ears, mouth, and nose, and all of that can end up lowering worker productivity.

Distribution on the vine. Leafhoppers overwinter as adults, and in spring they immediately begin feeding as soon as there is green grape leaf tissue to feed on. All early season injury shows up on the first basal leaves, and this is where the first-generation eggs are laid. First-generation nymphs are found almost exclusively on the basal leaves, whereas most second-generation nymphs are found on leaves from about node four to node eight, and third-generation nymphs can be found just about anywhere on the canopy except on immature leaves. The vast majority of nymphs and adults are found on the leaves' lower surface.

Regional distribution. WGLH occurs throughout California, Oregon, and Washington. In the eastern half of the San Joaquin Valley its numbers are highest early in the season and decline after that because of competition from VLH. In the North Coast and Central Coast areas and the west side of the San Joaquin Valley, WGLH is the primary leafhopper pest. In Arizona, the Coachella Valley, the Temecula Valley, the east side of the San Joaquin Valley, the lower Sacramento Valley, and the warmest regions of the North Coast, VLH is the primary leafhopper pest.

Development. After feeding on grape leaf tissue for a week or two in the spring, the overwintering generation will mate, and about 2 weeks later the females will lay eggs. Of the two species, WGLH has a lower developmental temperature threshold, so it hatches out earlier than VLH if both species are present. Each species develops through five instars before molting to adulthood. In Winkler Regions I and II, there are two generations per season, Regions III and IV might have a full or partial third generation, depending on the weather, and Region V has three distinct generations.

Sampling and EIL. The common practice is to visually inspect the leaves and count immature

Figure 9.25. Fifth instar variegated leafhopper. *Photo* by Michael Costello.

Figure 9.26. Adult variegated leafhopper. *Photo* by Michael Costello.

Figure 9.27. Leafhopper leaf injury. *Photo* by Michael Costello.

leafhoppers. Newly hatched nymphs are visible to the naked eye, but because they are so small, they are best counted with the aid of a 10× hand lens. The second to fifth instars are easy to see without a hand lens.

Select leaves from nodes on the cane based on leafhopper distribution on the vine; that is, the first generation should be sampled on leaves from nodes one through three and the second generation starting with node four and continuing to the most recently matured leaf. The third generation is rarely sampled, since chemical treatments based on nymphal density are seldom made that late in the season; instead, chemicals are directed at adult knockdown. Leaves should be selected in an unbiased manner (i.e., do not limit your samples to the leaves on which you see white stippling). In most regions, sampling should begin by mid-May, although in a year with a warm spring, nymphs may appear before then. Samples should be taken weekly, with enough leaves examined to give a precise estimate, keeping in mind the reasonable upper limit of about 30 samples per block see "Sampling (monitoring)" in the "Integrated Pest Management" section earlier in this chapter).

The EIL for leafhoppers is really a moving target, depending as it does on a number of variables that include (1) grape commodity type (i.e., wine, raisin or table [table grapes have a lower EIL because of the honeydew-sooty mold quality issue]), (2) canopy density and management (more leafhopper injury can be tolerated if there is a dense canopy, while just the opposite is true if you practice heavy leaf thinning), (3) harvest method (i.e., machine or manual [worker annoyance is not as great a concern for machine-harvested vines]), and (4) grower tolerance for leafhopper presence or damage. In general, a conservative EIL for table grapes and coastal winegrapes is a peak density average of about 2 to 4 nymphs per leaf, and for raisin grapes and Central Valley winegrapes an average of about 8 to 10 nymphs per leaf. Of course, these numbers may be increased or decreased depending on grower experience and tolerance level.

Cultural controls. Leafhoppers are sensitive to the water status of the vine, and as such, judicious use of irrigation can help you to manage them. Leafhopper nymphal density is lower on vines that are irrigated less, and evidence points to an explanation of lower leafhopper oviposition rates in leaves that are somewhat water stressed. What is not known

is the exact level of water tension that discourages oviposition. We do know, however, that an extreme water deficit can exacerbate other arthropod pest problems such as Pacific spider mite, so it is a good idea to keep water stress from exceeding a leaf water potential of about 16 bar (−1.6 Mpa), as measured using a bagged leaf, for any length of time. Initiation of water stress is not recommended before berry set, yet successful decreases of second-generation nymphal density have been achieved with as few as 3 weeks of post–berry set water stress at 50 percent of the crop's actual evapotranspiration (ET_c). Mid- to late-season irrigation in excess of 80 percent of ET_c would probably encourage shoot growth that would attract female leafhoppers looking for oviposition sites. Realize also that water deficits below 80 percent of ET_c are likely to decrease your yield.

Leaf removal, a common practice in many coastal winegrape and Coachella and San Joaquin Valley table grape vineyards, can be applied to leafhopper management. This practice, which increases air circulation around the clusters, exposes berries to more light, and improves disease management (see chapter 8), can be timed to correspond to the peak of first-generation nymphal density. Because most first-generation nymphs are on the basal leaves, you can help lower the population by pulling these leaves. However, timing is critical, and if this practice is to be successful it is imperative that you take weekly samples and record their numbers so you will be able to track leafhopper density and know when is the best time for this practice.

Cover cropping can be used to manage leafhoppers, but primarily through the mechanism of water status management. A permanent ground cover that stays green all season will compete with the vines for water and make the vines less attractive to leafhoppers. This is a more common practice in the San Joaquin Valley vineyards using flood or furrow irrigation. Cover crop management of this type will usually reduce vine vigor (and yield) and have an effect on leafhoppers similar to deficit irrigation. A noncompetitive ground cover such as an unirrigated but mowed annual cover crop or a perennial native grass will probably have little effect on vine vigor and, subsequently, leafhopper density. There is little evidence to indicate that a cover crop or summer ground cover encourages more natural enemies that feed on leafhoppers, but even if that were the case, it would probably not be as significant a control factor as water management.

Sticky yellow tape (Figure 9.28) has been used with some success, but it too has its drawbacks. For one, its use is fairly labor intensive. A common practice has been to put the tape out just before budbreak, placing it on the outside borders of a block and then at every 10th or 15th row. Overwintering adults making their way from alternate vegetation sources and those moving from vine to vine become caught on the tape. In theory, this could have a high impact on the leafhopper population, judging from the thousands of adult leafhoppers that are often captured on the tape. Another downside is that the tape only traps the overwintering generation of leafhoppers, so it only impacts the first generation for the season. After completion of the first generation, adults from elsewhere could migrate in and lay their eggs for the second generation. Another concern is the great amount of sticky tape waste that has to be rolled up and discarded when it is no longer useful—and the current product is not recyclable.

Biological controls. The most important natural enemies of leafhoppers are parasitic wasps of the *Anagrus* genus that lay their eggs into leafhopper eggs. The immature wasp develops within the leafhopper egg, consuming its contents. In some regions, *Anagrus* spp. can be very effective against WGLH, with high parasitism rates in the first and second generations. In other regions they seem to be less effective, and differences in insecticide use alone cannot explain the discrepancies. Recent

research has found that sulfur is toxic to *Anagrus erythroneurae* Trjapitsyn and Chiapini, but it did not appear to affect the wasp's performance in the field. Leafhoppers overwinter as adults, yet *Anagrus* spp. can only overwinter in an egg.

Attempts have been made to supplement the vineyard environs with plants that provide habitat for *Anagrus* spp.—plants that are hosts to leafhopper species that overwinter in the egg stage. One initial test used blackberries (*Rubus* spp.), which host blackberry leafhopper (*Dikrella californica* [Lawson]), but this was unsuccessful, presumably because blackberries were not suited to the conditions of a cultivated vineyard. Subsequently, French prunes (*Prunus domestica*), which support prune leafhopper (*Edwardsiana prunicola* [Edwards]), were planted around and sometimes within the vineyard. This also failed. Unpublished data by Kent Daane (University of California, Berkeley) suggests that there is simply too little prune vegetation compared to grape vegetation for the prune tree to sustain a large enough population of prune leafhoppers to make a difference. The same logic may well explain the earlier failure of the blackberry test. In the fall, when literally millions of adult *Anagrus* females are seeking eggs in which to lay their eggs, they quickly find the relatively small supply of alternate leafhopper eggs, effectively eliminating the population of those leafhoppers from the surrounding host vegetation (e.g., French prunes or blackberries). The only situation in which this strategy can be successful, then, is when the proportion of grapes to alternate vegetation is nearly equivalent. For the vast majority of growers, such an arrangement simply is not economical.

Spiders make up the only other significant group of natural enemies that feed on leafhoppers. Some of the more common species observed feeding on leafhoppers are the agrarian, or yellow sac, spiders *Cheiracanthium* spp. (Figure 9.1), the corrinid sac spider *Trachelas pacificus* (Figure 9.2), cobweb weavers (*Theridion* spp.) (Figure 9.3), and jumping spiders (many species). While it is considered beneficial to have a healthy community of vineyard spiders, just how much of a contribution they make to leafhopper management is not known, and a grower cannot count on them to maintain leafhopper densities below the EIL.

Figure 9.28. Sticky yellow tape. *Photo* by Michael Costello.

Chemical controls. A few chemical options are available to organic growers for leafhopper management. Chemical controls, and especially the contact materials (oils and soap), are probably most effective for the first generation, since the relative lack of foliage on the vines allows better spray coverage. However, if you can achieve good coverage at midseason, all of these materials can be effective against the second and even third generations. The difficulty with treating later generations is that the eggs do not hatch all at once, so two, or three sprays at 5- to 7-day intervals may be needed.

Oils can be effective (see the earlier discussion of oils in the "Spider Mites" section). For leafhopper management, oil applications need to be precisely timed to target the newly hatched nymphs. Mineral and vegetable oils will be most effective against nymphs in the first instar and less effective against the second instar, although a larger droplet size might improve efficacy. Third, fourth, and fifth instars are difficult to kill with oils. Essential oils have not been adequately tested against WGLH and VLH.

Soap can also be effective, but, as with oil, it is more effective if applied against the earliest leafhopper instars. Soap works by interacting with the waxy cuticle and subjecting the leafhopper to desiccation. Make sure to use insecticidal soap that is OMNI approved.

Botanicals such as pyrethrum and neem are also effective against leafhoppers, although they must be used with great caution to prevent the development of resistance. Neem is an insect growth regulator that acts as a chitin synthesis inhibitor. Its application must be timed properly, as it is most effective on the earliest instars.

Kaolin is a very fine clay that repels, deters, and annoys leafhoppers. It requires multiple applications in order to be effective, and at this point most grape growers who use it do so for control of sharpshooters rather than *Erythroneura* spp.

Sharpshooters

There are four species of sharpshooters in California that can impact vineyards. All of them have the potential to carry the bacterium *Xylella fastidiosa*, which causes Pierce's disease, a fatal affliction of *V. vinifera*. Those native to the West Coast are the blue-green sharpshooter (BGSS, *Graphocephala atropunctata* [Signoret]) (Figure 9.29), the green sharpshooter (GSS, *Draeculacephala minerva* Ball)

(Figure 9.30), and the red-headed sharpshooter (RHSS, *Carneocephala fulgida* Nottingham) (Figure 9.31). The fourth is the glassy-winged sharpshooter (GWSS, *Homalodisca vitripennis* [Germar] (Figure 9.32) [formerly *H. coagulata*]), a native of the southeastern United States.

Description. All sharpshooters have the general look of a leafhopper but with a typically more pointed head and a protruding face region. Nymphs of the three native sharpshooters have pale coloration; in other respects, they resemble adults. GWSS nymphs are a dark gray. BGSS, GSS, and RHSS nymphs are about 0.3 inch (7.6 mm) long, whereas GWSS measures about 0.5 to 0.6 inch (12.7 to 15.2 mm) long. The adult BGSS head is blue-green and relatively blunt; its blue-green wing covers have dark striations. The GSS is a uniform kelly green in summer and a dull brown in winter and has a distinctively pointed head. The RHSS is pale green with a moderately pointed, reddish head. The GWSS is uniformly earth-brown, with very large eyes and a large, bulging face region.

Because it only takes a single sharpshooter to inoculate many vines with *X. fastidiosa* bacteria, efforts at controlling the insects must be widespread, preventive (i.e., applied before sharpshooters are observed in the vineyard) rather than reactive, and virtually 100 percent effective, making it an expensive treatment that has the potential to disrupt other existing biological controls.

Injury and damage. Sharpshooters feed on a plant's water- and mineral-conducting tissues (xylem). It would take an extremely high density of sharpshooters, though, to cause injury such as shoot stunting or water stress. However, all sharpshooters have the potential to carry *X. fastidiosa* bacteria, which can colonize the xylem and clog it with their sheer numbers and secretions. This can cause leaf scorch, water stress, and, ultimately, vine death.

Distribution on the vine. BGSS, GSS, and RHSS feed on the grapevine's shoot tips. It is very unusual to observe GSS or RHSS actually feeding in a vineyard, since grape is not one of their preferred hosts, and they do not linger in the vineyard; however, BGSS will persist and even reproduce on grape. GWSS feeds anywhere on grape shoots, even on 1-year-old wood, and it can also reproduce on grape.

Regional distribution. The BGSS is found primarily on the coast, and even though it is generally associated with riparian vegetation it is also

Figure 9.29. Blue-green sharpshooter. *Photo* by Jack Kelly Clark.

Figure 9.31. Red-headed sharpshooter. *Photo* by Jack Kelly Clark.

Figure 9.30. Green sharpshooter. *Photo* by Jack Kelly Clark.

Figure 9.32. Glassy-winged sharpshooter. *Photo* by Jack Kelly Clark.

common on ornamental and landscaping vegetation. GSS and RHSS are found throughout the state and feed primarily on grass. GSS is the more common of the two. GWSS is found throughout Southern California, and at the time of this writing it was established in pockets of several Central Valley and Bay Area Counties.

Development. Sharpshooters overwinter as adults and are active during warm winter days. In spring they may migrate onto grapevines. The BGSS feeds readily on grape shoot tips, where it lays its eggs to produce the next generation. BGSS only has one generation per year. GSS and RHSS migrate from nearby grasses and feed for short periods on grape shoots, but they do not stay long. They have several generations per year. The GWSS has the ability to switch host plants rapidly and can migrate over relatively long distances. It is thought to be able to fly up to 400 meters (¼ mile) at a time. It has two

generations per year in Southern California.

Sampling and EIL. The most common way to sample for adult sharpshooters is to set out sticky yellow cards on the edges of the vineyard, especially in areas bordering riparian or pasture lands. Check the cards about once a week. There is no estimated EIL for any of the sharpshooters, and the decision to apply an insecticide is typically based on the degree of damage local vineyards have sustained from Pierce's disease in the past. The presence of at least one sharpshooter on a card is usually used as a signal to apply an insecticide treatment. Bear in mind, however, that early season populations of BGSS, GSS, and RHSS are potentially more damaging than those in the mid to late season. This is because BGSS, GSS, and RHSS feed on the shoot tips, so the bacteria transmitted by these species take a long time to reach the head of the vine. The same is not true for infections from GWSS, which can feed farther down the vine. Mid to late season Pierce's disease infections from BGSS, GSS, and RHSS only affect the upper half of the cane and, at least on spur-pruned vines, those areas will be pruned off in the winter, except, of course, in young vineyards that are still in the training phase. For these sharpshooter species, then, chemical treatment of a mature vineyard is only an early- to midseason consideration and, for the most part, one that would only need to be applied to parts of the vineyard that border on riparian or grassy areas.

Cultural controls. As mentioned earlier, the ultimate control for Pierce's disease is management of the disease itself, not its insect vector (see chapter 8). As far as insect management goes, the most effective cultural controls are those that attempt to keep sharpshooters out of the vineyard. This is a difficult undertaking and may only slow the invasion. Still, there have been attempts to plant non host vegetation (non host for Xylella as well as for sharpshooters) such as redwoods between vineyards and riparian or other areas. In vineyards near riparian areas, one study showed that by removing plants that are reservoirs of Xylella and replacing them with other native plants that are not good hosts, a grower can lower the incidence of disease. However, this is an expensive, time-consuming, and long-term proposition, as you are likely to need permits from agencies such as the California Department of Fish and Game and the U.S. Fish and Wildlife Service. Removal of grasses that serve as habitat for green and red-headed sharpshooters may help to reduce incidence of these insects. Host grasses may be weeds within the vineyard, or they may be growing in nearby riparian areas or pastures. If glassy-winged sharpshooter is established in your region, do not grow grapes near a citrus orchard, as citrus is a favorite host plant. Attempts to keep GWSS out of vineyards by erecting a 20-foot-high fence or netting have met with some success. GWSS generally flies in a straight line some 5 to 10 feet off of the ground. Still, when GWSS density is high, some individuals are almost certain to make it over even the highest fence.

The only other possible cultural control for sharpshooters is water management. Sharpshooters are obligate xylem feeders, so the more water stress a plant is under, the harder it is for sharpshooters to extract the xylem fluid. They are known to avoid plants with low water potential. The degree of water stress needed to restrict sharpshooter feeding is not known. Even if that information were known, it

would be difficult to eliminate sharpshooter feeding solely through water management, since (except in desert regions), a grower simply cannot control a vine's water status throughout the entire season. Winter and spring rainfall in most regions and in most years make it impossible.

Biological controls. The most important biological control agents for sharpshooters are egg parasitoids, tiny wasps in the genus Gonatocerus (family Mymaridae). These parasitoids reduce populations of all sharpshooter species and are undoubtedly an important element in overall control. However, no parasitoid can provide 100 percent control, and it only takes a few sharpshooters infected with Xylella to cause a lot of damage to a vineyard. Moreover, G. ashmeadii is more effective on the second generation of GWSS. Attempts are being made to find and import parasitoids that have a greater impact on the first generation. Of the generalist predators, earwigs and lacewings are known to feed on GWSS egg masses.

Chemical controls. A general lack of OMNI-approved systemic chemical controls for sharpshooters makes this avenue of control difficult for organic growers. Soap and oils, as contact materials, are not effective on adult sharpshooters. Botanicals would have to be applied very frequently (at least once a week) to obtain the necessary control. Kaolin is a very effective repellent for sharpshooters and can be applied to vines along the interface with riparian or grassy areas, but you have to apply it often enough to create multiple layers and to cover new growth. In areas infested with GWSS, growers have used kaolin to treat neighboring host vegetation, including citrus. A regional organic plan for sharpshooter and Pierce's disease management would be challenging to establish and implement, but it would be possible.

Mealybugs

Mealybugs are in the same order (Hemiptera) as leafhoppers and sharpshooters, and they fall into their own family, the Pseudococcidae. There are now four mealybug species that can reach pest status in California vineyards: grape mealybug (GMB, *Pseudococcus maritimus* [Ehrhorn]) (Figure 9.33), obscure mealybug (OMB, *Pseudococcus viburni* [Signoret]) (Figure 9.34), longtailed mealybug (LTMB, *Pseudococcus longispinus* [Targioni-Tozzeti]) (Figure 9.35), and vine mealybug (VMB, *Planococcus ficus* [Signoret]) (Figure 9.36). Of these, GMB and LTMB have been in California for as long as grapes have been cultivated here, and perhaps longer. OMB was first observed on the Central Coast in the 1980s, and VMB arrived in the Coachella Valley in the early 1990s. Mealybugs are unusual among grape arthropod pests in that their infestation is greatly influenced by the presence of ants.

Description. Mealybugs are oval insects and are wingless with the exception of adult males, which are not commonly seen and contribute very little to the actual damage done to the grape crop. Mealybugs have a pinkish body covered with white, waxy extensions of varying shapes and lengths. The adult females of all species are about $1/10$ to $1/5$ inch (2.5 to 5 mm) long. LTMB has two caudal filaments (tails) that extend more than three-quarters of the body length, distinguishing it from GMB and OMB, whose tails are between one-quarter and one-half of its body length. GMB and OMB are very difficult to tell apart on any basis other than minor characteristics that can only be discerned by experts. Deliver specimens for identification to a farm advisor at your local UC Cooperative Extension county office or to your county agricultural commissioner's office. Alternatively, you can make an educated guess while you wait for the results. Because GMB goes into diapause during the winter, its generations during the other seasons are well synchronized, and only one or two stages (egg mass, any of three instars, and adult) are present at any given time. In contrast, OMB never goes into diapause so all of its stages are present in overlapping generations all year round. The VMB is easy to identify: it has numerous short, waxy extensions that circumscribe its body, and it has no tails.

All of the mealybugs discussed here lay 100 to 300 eggs within a mass of white, woolly wax, with the exception of LTMB, which gives live birth. First instar mealybugs, called "crawlers," have no waxy covering. There are three instars before mealybugs transform into the adult stage. Males also have a pupal stage that transforms it into a small, winged individual. With each molt, the mealybug casts off its skin layer (or cuticle), leaving the newly molted mealybug pink.

A note on ants: The presence of ants factors significantly in the degree of mealybug infestation and injury. In coastal vineyards, the primary (and often only) ant species found is the Argentine ant,

Figure 9.33. Grape mealybug. *Photo* by Jack Kelly Clark.

Figure 9.34. Obscure mealybug. *Photo* by Michael Costello.

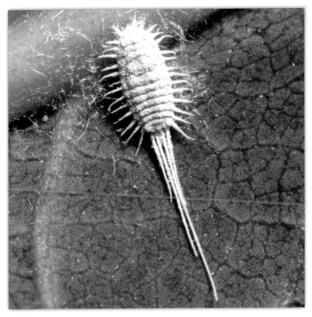

Figure 9.35. Longtailed mealybug. *Photo* by David Rosen.

Figure 9.36. Vine mealybug. *Photo* by Kent M. Daane.

Linepithema humile (Mayr). This is the small (1/8 inch [3 mm] long), dull brown ant that commonly invades kitchens, garbage cans, gardens and commercial agricultural fields en masse. In California's Central Valley, the native gray ant, or field ant, (*Formica aerata* Francoeur) is found alongside the Argentine ant. The native gray ant is larger (1/3 inch [7 mm] long), has a grayish brown abdomen, and the ants forage individually rather than as a group.

Injury and damage. Mealybugs feed on plant sap (phloem) and produce a sticky honeydew as a waste product. The honeydew is a fruit contaminant and the main injury concern for mealybugs. Prolonged feeding by mealybugs on leaves can cause discoloration similar to that caused by Willamette spider mite, but this particular symptom typically affects only the basal leaves, contributing little or nothing to economic damage. The main problem occurs when mealybugs migrate to the clusters after veraison and foul them with their honeydew, cast skins, and egg masses (Figure 9.37). Heavy infestations can lower berry size and reduce cluster weight. By far the worst damage is done by VMB, which produces copious amounts of very thick honeydew. VMB can devigorate vines to the point that they become economically unproductive. All of the mealybug species are potential vectors for the leafroll viruses, although they are not thought to be important in the spread of the disease.

Distribution on the vine. Mealybugs overwinter as eggs and crawlers under the vine's bark, with the exception of OMB, which can be found in all stages all year round. The VMB overwinters on grape roots, but can also be found on vine roots in the summer. In spring, the crawlers migrate from their overwintering location to newly developing leaves, where they feed for a number of weeks. Then they migrate back to the bark (or roots in the case of VMB) to mate and lay eggs, and the new generation of crawlers migrates back out to the canopy to feed on leaves, shoots, and clusters. Mealybugs tend to infest clusters that touch the wood of the cordon or trunk, and they are often found on individual berries that come into contact with wood.

Regional distribution. GMB can be found throughout California and in Oregon and Washington, whereas LTMB is primarily found on California's Central Coast. GMB was formerly the most important pest in the cooler areas of the Central Coast but has more recently been supplanted by OMB. GMB is still the most important mealybug pest on the North Coast and in the San Joaquin Valley. The VMB is currently established in 16 California counties where grapes are grown commercially. Most of its spread has been via nursery stock, so its presence in any given region is spotty. VMB is likely to replace GMB and OMB in importance as it continues its spread through California.

Development. Whereas GMB and LTMB have two generations per year, OMB may have as many as three. Crawlers migrate to new leaves in the spring, return to the bark in early summer (June), and migrate back to the leaves, shoots, and clusters in midsummer (mid-June to early July). VMB may have as many as seven generations in a year, depending on the region.

Figure 9.37. Mealybug mess. *Photo* by Larry L. Strand.

Sampling and EIL. Mealybugs are particularly challenging to sample because they are so well hidden. To get a general idea of the level of infestation, you can peel back the bark on a vine's trunk to reveal adult females and egg sacs in winter or at midsummer.

In the delayed dormant to budbreak period (early to late March), you will need to sample to determine when crawlers are active, which will determine the optimum timing for insecticide applications. Then you can generally use a nominal EIL; that is, the treatment decision is based largely on the vineyard's history of infestation and your PCA's experience, and little effort is made to quantify the samples. Alternatively, you can use an EIL similar to what is recommended on the UC IPM website (http://www. ipm.ucdavis.edu/): spray if 20 percent of the spurs you inspect have crawlers, or 10 percent if you are growing table grapes or high-value winegrapes. Because OMB crawlers can emerge at any time over a period of weeks, weekly sampling may be necessary, along with multiple chemical applications. You can monitor by means of direct visual inspection of spurs, but that is time consuming and the crawlers can be hard to find. PCAs have had some success using double-sided sticky tape wrapped around canes or spurs. These improvised traps can be placed out and retrieved weekly, and you can count the trapped crawlers with a hand lens or microscope. The tape must be reapplied after a rain. You can use the same procedures to sample for second-generation crawlers from late May to early June.

In midsummer (June and July) you can collect samples from clusters. The most efficient method is to limit your sampling efforts to clusters that are touching the cordon or trunk, since these are the clusters that mealybugs prefer. Sampling just checks for presence or absence, so it only takes a few seconds to visually scan the berries. Infestations are notoriously spotty, so you may have to look at as many as 100 clusters. The EIL at this point in the season is not well defined, so once again the treatment decision is largely based on the grower's or PCA's tolerance level and experience. The usual tolerance level for mealybug infestations on table grapes and high-value winegrapes is very low; a 1 percent infestation would be treated in most cases. For raisins and low- to moderate-value winegrapes, a higher percentage is tolerable, perhaps as high as 10 percent. The trouble is, mealybugs are well hidden in midsummer, and they are difficult to kill with conventional insecticides, much less with the less-potent, organically approved materials.

In areas where VMB is not established, you can sample for it using pheromone traps. The trap's lure emits a synthetic sex pheromone that attracts male mealybugs. Lures will last for about 3 months. If you find one or more male mealybugs in a trap, take it to your local UC Cooperative Extension county office or the county Department of Agriculture for identification. If the specimen is determined to be VMB, a ground crew should undertake direct visual inspection, find the infestation, and destroy it. If only a few vines are affected, it may be prudent to pull them out altogether, bag them, and burn or bury them at a location far removed from any vineyard.

Cultural controls. Mealybugs really like hiding places. Some growers have gone so far as to strip the vines of all bark in the winter in order to eliminate overwintering habitat, but this is expensive, probably not very healthy for the vine, and very ecologically disruptive.

Because the first infected clusters tend to be those that touch the wood, any training and pruning methods that encourage free-hanging clusters and prevent them from touching wood can definitely help. Cane-pruned vines are less susceptible because there is no cordon for clusters to hang down upon. Spur-pruned vines are much trickier because the spurs need to be positioned as near to right angles from the cordon as possible, and perhaps pruned longer (leaving a three-bud spur as opposed to a two-bud spur). This can be a very difficult arrangement to manage and maintain. Another option is midsummer removal of those clusters that touch the wood, but that is an expensive, labor-intensive operation.

Grape variety can play a role in the degree of mealybug infestation. The longer the mealybugs are in the clusters, the more the cumulative damage. Early ripening winegrape varieties, therefore, are in theory less vulnerable, although overlapping generations of OMB on the Central Coast allow plenty of time for infestation of all varieties, no matter how early they mature. Early table grape varieties are likewise less susceptible, although, given the low EIL for mealybugs on table grapes, a variety in the Central Valley would have to mature by mid-July in order to beat the migration of second generation mealybugs.

Ant control is an indirect form of mealybug control, since ants tend mealybugs for their honeydew supply and protect them from their natural enemies. Sticky substances applied in a band around the trunk are very effective ant barriers, so long as there is absolutely no path left uncovered. However, a sticky band is expensive to apply and must also be maintained. Some of the sticky materials are designed to last longer in warm weather, but dust and dead ant bodies will eventually render any such barrier ineffective, no matter how viscous the material may be.

Although extreme, the destruction of vines in areas where VMB is not yet fully established is an effective cultural control method (see procedures described earlier, under "Sampling and EIL"). Until biological controls improve, VMB will continue to be very difficult if not impossible to manage organically. If the pest is allowed to establish, the only economically sound option for its control may be the use of conventional insecticides. If you want to maintain organic certification, you may find it is worth your while to remove a small portion of your vineyard in order to keep that status.

Biological controls. The GMB has the most extensive complex of biological control agents of the four mealybug species. No fewer than six species of parasitic Hymenoptera (wasps) attack it. The most common are *Acerophagous notativentris* (Girault) and *Pseudaphycus angelicus* (Howard). If unhindered by insecticides and ants, these parasites usually keep grape mealybug numbers very low. In the 1990s, several hymenopteran parasitoids were imported from Chile to help control OMB, and the parasitoids *Pseudaphycus flavidulus* (Brethes) and *Leptomastix epona* (Walker) were established in California. However, the Argentine ant very aggressively attacks these wasps and is responsible for the low percentage of parasitism in most Central Coast vineyards. The VMB is commonly attacked by *Anagyrus pseudococci* (Girault), which is also the most important parasite of VMB's close relative, citrus mealybug. VMB parasitism can reach high levels late in the season but usually only after any major damage has already occurred.

A number of predators feed on mealybugs, including lacewings and *Cryptolaemus montrouzieri* (Mulsant), the mealybug destroyer. *Cryptolaemus* is not very common in the Central Valley, apparently because of its sensitivity to cold winters. It is also rare in vineyards with mild climates, such as the Edna Valley in San Luis Obispo County, although

curiously, it is quite common on citrus in that area. Lacewings are known to feed readily on mealybugs and can be found in higher numbers in vineyards infested with mealybugs, but as yet there is little evidence with regard to the impact of lacewings on the four mealybug species that attack grape. Lacewing eggs are available for purchase and release, but if ants are present (as they often are) they will seek out and eat the lacewing eggs before they hatch.

In theory, cover cropping can help a grower manage mealybugs, but the effect is probably only indirect. One study showed that a cover crop of common vetch, which has exposed (extra-floral) and readily available nectaries, increased GMB parasitism. Presumably, nectar from the common vetch drew ants away from the vines, allowing parasitoids to attack the mealybugs without any interference from the ants.

Chemical controls. For organic growers, the only times that it makes sense to spray for mealybugs are before budbreak and early in the spring to target the crawler stage, or, at the very latest, in midspring to target the early instars on leaves. When targeting crawlers or early instars, you may need to make multiple applications. Spray coverage needs to be especially good if you are targeting the later instars, since the individuals are larger and less susceptible to chemical controls. Organic rules only allow lime sulfur as a dormant application. Oils, soap, and neem are effective against mealybugs. Pyrethrum should be used only as a last resort, because of its higher potential negative impact on natural enemies. Again, with any of these OMRI-approved materials it is difficult at best to control an infestation at midsummer or later, when the infestation has established itself in the clusters.

Bait stations for ant control may be a useful tool in a mealybug control effort and may provide the only way to achieve effective biological control in some areas. Unfortunately, at the time of this writing there are no active ingredients for ant baits registered for organic use in California vineyards. A bait station would need to be infused with a sugar-based bait to attract the Argentine ant, and this would also attract gray ants. Bait stations are easy to make (see Cooper and Daane in the "References/Resources" section). Initial costs for the placement and maintenance of bait stations are high in terms of materials and labor, with a suggested density of 15 to 20 bait stations per acre. However, research indicates that in the long term (beyond 3 years), the bait stations will reduce ant populations significantly so that over time you can reduce the number of bait stations per acre.

Lepidopteran Pests

Lepidoptera is the order name of the butterflies and moths, and in California grape production there are four moth species that are pests or potential pests and perhaps more on the way. Omnivorous leafroller (OLR, *Platynota stultana* Walsingham [family Tortricidae]) (Figure 9.38) and orange tortrix (OT, *Argyrotaenia citrana* [Fernald] [family Tortricidae]) (Figure 9.39) are ecologically similar in that their females lay eggs on young berries and their larvae (caterpillars) feed on developing fruit. In contrast, the larvae of grapeleaf folder (GLF, *Desmia funeralis* Hübner [family Pyralidae]) (Figure 9.40) roll leaves up and feed on them from within, and the larvae of grapeleaf skeleton-izer (GLS, *Harrisina brillians* Barnes and McDunnough [family Zygaenidae]) (Figure 9.41) feed on leaves and can strip them of all tissue down to the main veins. These latter two are typically kept under excellent biological control and only cause problems in unusual outbreak years or when natural enemies are disrupted by insecticides. In 2007, a fifth moth species, the light brown apple moth (*Epiphyas postvittana* [Walker] [family Tortricidae]) (Figure 9.42) native to Australia, invaded California. If it becomes established, treatment for it will be similar to that used for OLR and OT.

Description. OLR and OT moths are about 7 to 8 mm long, and their wings form a bell shape when at rest. OLR has a slightly longer "snout" (the labial palps) than OT. The first instar of each species is about 1/12 inch (2 mm) long and pale green with a dark head capsule. Fifth instar OLR has two cream-colored spots on the top of each thoracic segment, spots that are lacking in OT.

GLF moths are about ½ inch (12 mm) long with a wingspan of about 1 inch (25 mm) and are dark with white spots on fore- and hind wings. Larvae are about 1/16 inch (1.5 mm) long when newly hatched and are tan to light green in color. Late-instar larvae have a dark crescent-shaped mark on each side of the second thoracic segment. The fifth instar is about 4/5 inch (20 mm) long when fully grown. GLS moths are about 3/5 inch (15 mm) long with a wingspan of about 1.2 inch (30 mm) and a metallic blue-black color that makes them look almost wasp-like. First through third instars are cream colored, but later instars begin to take on a striking pattern

Figure 9.38. Omnivorous leaf roller larva. *Photo* by Michael Costello.

Figure 9.39. Orange tortrix larva. *Photo* by Jack Kelly Clark.

Figure 9.40. Leaf injury from grapeleaf folder larva. *Photo* by Jack Kelly Clark.

Figure 9.41. Grapeleaf skeletonizer damage. *Photo* by Michael Costello.

Figure 9.42. Light brown apple moth larva. *Photo* by Jack Kelly Clark.

of blue, black, and yellow rings. The fifth instar is about 3/5 inch (15 mm) long and has urticating (stinging) body hairs.

Injury and damage. OLR and OT feed on leaves, but the amount of leaf loss is minor and causes no real damage to the vine. There have been reports of bud injury just prior to budbreak, but this is very unusual and probably only occurs when an early biofix date (e.g., February 1 in the Central Valley) is coupled with a very high OLR or OT overwintering population. In early spring, larvae are primarily found on leaves, flower parts, and newly set berries; injury to the latter causes fruit scarring or causes individual berries to fall off but does not result in any economic loss. However, as sugar begins to accumulate in the berries after veraison, OLR and OT

feeding creates entry wound opportunities for bunch rotting fungi and bacteria. One infected berry can inoculate neighboring berries, ultimately causing the entire cluster to rot and collapse.

GLF larvae feed on leaves and roll them up. At a high pest population density, some 50 percent of the canopy can be affected, significantly reducing the vine's photosynthetic area and leading to fruit sunburn. Only at very high population densities do GLF larvae feed on berries.

GLS larvae feed on leaves and can strip them of all tissue save the main veins. At high population densities, they can remove virtually every bit of green tissue on a vine, shutting down its photosynthesis and preventing further berry development. Only after it has consumed all leaf tissue will GLS begin to feed on grape berries. Older larvae also cause indirect damage by interfering with hand picking operations, as their stinging hairs can annoy workers and reduce productivity.

Distribution on the vine. Any of these lepidopteran larvae can be found on any leaves, but typically only OLR and OT larvae are found in the clusters. Only when GLF or GLS density is extremely high will they move onto the clusters and begin feeding on berries.

Regional distribution. OLR can be found in the San Joaquin Valley and on the Central Coast, whereas OT is found on the North Coast and Central Coast. GLF is found in the San Joaquin Valley and eastern San Luis Obispo County, and GLS is present in the San Joaquin and Coachella Valleys.

Development. The OLR and OT overwinter as late-instar larvae on weeds and unharvested berry clusters (mummies). They do not go into diapause but rather develop slowly through the winter, eventually pupating wherever they happen to be. On the coast, moths emerge in late December or early January, mate, and lay eggs on weeds or grape tissue. In the San Joaquin Valley, OLR emerges from February 1 through March 15. Both moth species lay their eggs on young leaves and flower clusters, and their larvae knit the tissue together to protect themselves as they feed. The larvae develop through the five instars, pupate on the vine, and emerge again. The OLR has three fairly distinct generations per year in the San Joaquin Valley, whereas OLR and OT may have as many as five generations on the coast.

GLF overwinters in the pupal stage in leaf litter or under the grapevine's bark. In spring, moths emerge and mate, and the females lay eggs singly on leaf tissue. Young larvae feed in protected areas where leaves are touching. Fourth instar larvae roll leaves up and then feed inside until they pupate.

GLS overwinters in the pupal stage under the bark and emerges shortly after budbreak. After mating, females lay clusters of eggs on leaves, sometimes just a few and sometimes as many as 300. Early instar larvae feed gregariously, each larva juxtaposed to another. After the third instar, larvae disperse and feed separately. Fifth instars crawl down to the trunk to spin a cocoon and pupate under the bark.

Sampling and EIL. Monitoring with pheromone traps is the most common sampling method for OLR and OT. Those traps do nothing, however, to help estimate pest density on the vines; rather, they help growers determine spray timing. In many areas, natural enemies are plentiful enough to provide control, so insecticide applications there are the exception rather than the norm. The decision to spray is often based on the area's history of infestation and the PCA's own experience (i.e., a nominal EIL), but these factors should also be coupled with direct visual inspection to estimate the degree of infestation. In areas that are chronically affected by these pests, growers should purchase a pheromone kit for OLR or OT (or both). The kit consists of a wax-coated cardboard trap with a sticky bottom and removable top, a wire hanger, and a lure impregnated with synthetic pheromone (Figure 9.43). The pheromone is slowly released over time (lures are rated for 30, 60, or 90 days). In coastal areas, place traps by January 1; in the San Joaquin Valley, they should be out by February 1. The pheromone attracts male moths in the area. When males are caught in the trap, you can assume that other males are in the area as well and that mating is taking place. The date of the first consistent catch of male moths is the biofix date that you will use for subsequent calculation of cumulative degree-days. Degree-days can be calculated based on readings from an on-site analog or digital thermometer, from commercial weather station networks, or from public weather stations (CIMIS and PestCast) accessible through the UC IPM website (http://www.ipm.ucdavis.edu).

One way to get cumulative degree-days for your area is online, through the UC IPM website. Choose "Degree-Days," then OLR or OT, and select "Run Model." Select a county, enter the biofix date as the start, and your estimated treatment date as the end. If the end date is still in the future, there is no cause for worry; the database will estimate future degree-days based on the 30-year average. Next, select specific stations in the selected county, and finally, select "Calculate." The program will provide daily and cumulative degree-days through the estimated treatment date. For degree-day calculation, the program defaults to the single sine method with a horizontal cutoff. If you use on-site weather data, an easy way to calculate degree-days is the rectangular method. Most of the time this method differs from single sine by only a few percentage points. Record the maximum and minimum temperatures and then use the following formula:

$$[(max + min) \div 2] - LDT$$

where LDT is the lower developmental threshold; that is, the low temperature at which the insect goes into an arrested state. For OLR the LDT is 43°F (6°C) and for OT it is 48°F (8.8°C). Each day's degree-day total is added to the previous cumulative total. Studies suggest that the most effective treatment timing is between 700 and 900 degree-days °F (390 and 500 degree-days °C).

Good practice is to conduct direct visual inspection of clusters for OLR and OT starting at 500 to 700 degree-days °F (278 to 390 degree-days °C) after biofix. Sample a minimum of 100 clusters per block, drawing the samples from at least 25 vines. In spring, look for flower buds that are knitted together with webbing or berries that have been fed upon. With practice, you

Figure 9.43. Pheromone trap. *Photo* by Michael Costello.

should take no more than 10 to 15 seconds to analyze a cluster for injury or the presence of larvae. For the second generation, use the same methodology to look for berries knitted together and, again, feeding marks on the berries. The EIL is not well defined, but most growers, regardless of grape crop type, typically treat when 1 cluster in 100 is infested. Again, this will vary with the individual grower's tolerance and experience. The decision whether to treat for OLR or OT at the first or second generation is discussed below, under "Chemical controls."

Sampling for GLF and GLS is not regularly done, since these pests are usually under good biological control. On weekly visits to the vineyard, it is a good idea to keep an eye out for GLF leaf rolls, though. If they become quite numerous (say, 10% of the foliage is rolled up), you might be wise to consider treatment. For GLS, only a very unusual circumstance would call for chemical treatment. This species is attacked by a combination of tachinid fly and granulosis virus (see "Biological controls," below) which together keep the vast majority of GLS infestations small, localized, and short lived. If you do observe any infestation, make a note of its location and revisit that place every week; chances are very high that the biological control agents will act upon it and keep injury to a minimum.

Cultural controls. OLR and OT feed on a variety of alternate host plants, including weeds such as mare's tail (*Conyza* spp.), mallow (*Malva* spp.), mustard (*Brassica* spp.), and cultivated crops such as stone fruit and cotton. Although there have been no published studies to determine whether weed control will lower these pests' populations on the vines enough to prevent the need for a spray treatment, it is not a bad idea to keep the vineyard and surrounding areas free of alternate hosts, especially if the vineyard has a history of infestation. Also, since these pests feed on and overwinter in grape mummies, postharvest removal of unharvested clusters probably is a good practice, although, again, no formal studies have demonstrated whether it really makes a difference in OLR or OT density.

The planting of nectar-producing plants in hedgerows or other locations surrounding the vineyard may also help provide resources for lepidopteran natural enemies (see "Biological controls," below), although again, no formal studies have tracked parasitism rates or differences in OLR or OT infestation with and without a hedgerow.

Biological controls. Perhaps a dozen species of parasitic Hymenoptera (including those in the families Braconidae and Bethylidae) and Diptera (in the family Tachinidae) are known to attack OLR and OT, and in many areas they contribute greatly to overall control. The only candidates for natural enemy release are the *Trichogramma* spp. (family Trichogrammatidae), tiny, hymenopteran egg parasites.

No formal studies exist to test the effectiveness of *Trichogramma* releases against OLR or OT, but the practice has been studied on other pests and crops, such as codling moth in walnuts. In that case, it was estimated that a release of about 225,000 parasitized eggs per acre (about two packaged cards) lowered larval infestation from 4 percent down to 1 percent. A major barrier to the success of such a release is the presence of ants or earwigs, which will readily consume the eggs and the parasites they contain.

GLF is kept under control most years by a parasitic Hymenoptera called *Bracon cushmani* (Muesbeck). The moth does, however, have occasional outbreak years. These are probably due to weather fluctuations; they cannot be explained by the use of broad-spectrum insecticides that would kill off the parasite. GLS is kept under control by a combination of a parasitic tachinid fly (*Ametadoria misella* [Wulp]) and a granulosis virus. In San Diego County it is also attacked by a parasitic Hymenoptera, *Apanteles harrisinae* Muesebeck. The vast majority of GLS outbreaks are eventually brought under control by these natural enemies, with often only a few acres affected. Chemical treatment of initial GLS outbreaks only serves to make their long-term control more difficult.

Spiders make up the only other group of natural enemies with the potential to impact lepidopteran pests. Research indicates that the agrarian or yellow sac spider (*Cheiracanthium inclusum* [Hentz]) and a corinnid sac spider (*Trachelas pacificus* Chamberlin & Ivie) readily consume OLR larvae in the laboratory and the field.

As mentioned above, many of these natural enemies, including those in the hymenopteran families Braconidae and Bethylidae and the dipteran family Tachinidae, will supplement their regular feeding with other vineyard sources, such as floral nectar. This means that plants in or near the vineyard that produce nectar or support alternate prey can, theoretically, increase the staying time or longevity of these natural enemies. In addition, alternate plants may support populations of nonpest Lepidoptera that can be alternate food sources for these parasitoids. It should be noted, however, that studies on the benefits of alternate vegetation are mixed. Even when a natural enemy's population density is increased by the presence of floral food sources, results often show no realized effect on pest density. Though this does not mean that there is no effect, it does indicate that success is far from certain.

One way to conserve natural enemies is to avoid the use of broad-spectrum pesticides. Soap, oils, and neem have only a minimal effect on adult natural enemies, but botanicals such as pyrethrum are indiscriminate in their activity and should be used only under outbreak conditions.

Chemical controls. The most effective type of chemical control for organic growers is pheromone confusion, the principle of which is to flood a vineyard or orchard with sex pheromone so that male and female moths are unable to find each other. To date, the only pheromone system available is for OLR, and only the hand-placed pheromone-impregnated strips are OMRI approved. The sprayable (flowable) pheromone is not allowed.

There is no point in trying to use soap or oils for lepidopteran pests, as the application timing would have to be perfectly matched to first instar emergence. This is not only a difficult undertaking on its own, but because hatch does not happen simultaneously for all individuals it would require not one but a series of perfectly timed applications. Caterpillars quickly become less susceptible to soaps and oils as they get older. In addition, once OLR, OT, and GLF larvae begin to feed, they also start to cover themselves with silk or plant material, making them less exposed to these contact materials.

Use botanical insecticides such as pyrethrum with caution against these lepidopteran pests, as there is a great risk that you will disrupt the population of biological control agents. The only circumstance that would justify use of a botanical insecticide is a late-season outbreak of one of the lepidopteran pests.

Various formulations of *Bacillus thuringiensis* (Bt) are appropriate for lepidopteran pest control. Bt is a naturally occurring soil bacterium that produces a protein crystal toxic to immature Lepidoptera, Diptera, or Coleoptera, depending on the Bt strain. It is a "stomach poison" (really, a midgut disruptor), which means that the caterpillars have to ingest the

crystal in order for it to be effective. Another option is spinosad, a derivative of the bacterium *Saccharopolyspora spinosa* Mertz and Yao. Spinosad is a synaptic poison that activates a nerve cell's acetylcholine receptors. It can enter an insect by contact or ingestion. For both Bt and spinosad, adequate spray coverage and precise timing are very important. Both materials are most effective against young larvae and both break down quickly in the environment. Larval death is not instantaneous, sometimes taking several days.

In chronic problem areas for OLR (e.g., Caruthers, Fresno County) or OT (e.g., Soledad, Monterey County), the use of Bt or spinosad timed to the development model described earlier is recommended. An experienced PCA should decide whether to treat the first or the second generation. Each choice has its advantages. The advantage to treating the first generation is that berry clusters are small and open, making good coverage easier. Also, effective treatment of the first generation will slow the buildup of the second generation. However, there are also advantages to choosing not to treat the first generation. First of all, first-generation caterpillars do not cause economic damage. The few berries that they feed upon will either scar over or fall off the vine. Second, a spray applied early in the spring may not provide control throughout the season. First-generation moths may move in from outside the vineyard, making treatment of the second generation necessary, regardless. Treatment of the second generation is usually sufficient to keep the population low until harvest.

Chemical treatment for GLF is not recommended unless it is an outbreak year. In that case, the only choice available to organic growers is to use a botanical such as pyrethrum. Even this will provide less than a full kill, since caterpillars that have already begun to roll leaves up will be very difficult to reach and affect. The very effective biological control combination of tachinid parasitoid and granulosis virus against GLS should generally avert the need for chemical treatment for this moth. When treatment does become necessary, however, Bt or spinosad should be very effective, as the caterpillars are very exposed until late in the fifth instar.

Phylloxera

Phylloxera (*Daktulosphaira vitifoliae* Fitch) is about the only insect pest for which successful and long-term management by organic and conventional growers is exactly the same: proper rootstock selection. Phylloxera are insects very closely related to aphids, but they fall into their own separate family (the Phylloxeridae). Grape phylloxera is specific to *Vitis* species, and the European grape (*V. vinifera*) is quite susceptible and will go into decline and usually die from an infestation. Grape phylloxera is native to the eastern United States, and all native American *Vitis* and *Muscadinia* species have some resistance to the insect. Historically, phylloxera has been a barrier to *V. vinifera* cultivation. It was a major cause of failed plantings by seventeenth- and eighteenth-century colonists in eastern North America. Phylloxera has also been responsible for major destruction of existing vineyards, beginning in Europe in the 1860s (see Campbell 2004 for more information). More recently, California suffered a major phylloxera outbreak from the late 1980s through the mid-1990s because of the failure of rootstock AXR1. So long as the grower has selected rootstocks of pure American parentage, phylloxera should not pose a problem.

Description. Phylloxera are somewhat smaller than aphids (adult females are not much more than ¹⁄₆₄ inch (0.5 mm) long (Figure 9.44) and so are difficult to see with the naked eye. Although various color morphs are known (ranging from dull brown to gray), most individuals (including the eggs) are a bright orange-yellow and quite distinct. Once the soil is washed off of them, they can readily be seen with a 10× hand lens. Adult females are wingless and sack-like in shape.

Injury and damage. Grape phylloxera in California are restricted to the vine's roots, where injury occurs, although the ultimate damage is death of the entire vine. They feed on small feeder roots, causing a swelling and distortion of the root's shape known as a "nodosity." This damage reduces the number of root growing points, which in turn limits the vine's capacity for water and nutrient uptake. Feeding on larger roots results in root swelling, cracking, and blackening, known as a "tuberosity," which ultimately causes the vine's death (Figure 9.45). Recent studies suggest that secondary fungal infections of damaged roots may be partly or largely responsible for the vine's actual decline and death. Aboveground symptoms are characterized by an overall decline, with leaf and shoot symptoms that include stunted shoots and necrotic leaves, not unlike those caused by nematode infestation.

Regional distribution. Grape phylloxera is distributed worldwide and is widespread in California, although it has not been reported from the South Coast. In California, only the wingless, root-infesting phase is known, although there have been reports of nursery invasions of the leaf-infesting phase (see "Development," below). The leaf-infesting phase is found in the eastern United States and has also been reported in Oregon. Phylloxera does not do well on sandy soils, and vineyards with only a very low clay content in the soil are typically not susceptible to infestation. Indeed, large areas of the San Joaquin Valley with own-rooted vineyards on loamy sands and sandy loams have never had problems with phylloxera.

Figure 9.44. Phylloxera colony. *Photo* by Michael Costello.

Figure 9.45. When phylloxera feed on larger roots, they cause a tuberosity, a condition that will kill the vine. *Photo* by Jack Kelly Clark.

Development. In California and Washington state, established phylloxera are biotypes that only undergo asexual reproduction and that remain only on the vine's roots. Nymphs hatch from eggs laid asexually by adult females. Nymphs develop through several instars before reaching adulthood as females, and the cycle repeats. In the eastern United States and Oregon, some nymphs will develop wingpads in midsummer and emerge from the soil as winged females. These begin a sexual phase that ultimately results in galls on the leaves of susceptible varieties (e.g. 'Concord' and French hybrids). Some of the gall-making females then migrate to the roots to begin a new phase of asexual reproduction.

Sampling. Growers do not routinely sample for phylloxera, only doing so as part of a trouble-shooting effort to analyze a general vine decline. Sampling requires root tissue, so you have to take a shovel and dig about 1 foot (0.3 m) out from the main trunk and 1 to 2 feet (0.3 to 0.6 m) deep. Look for sections of roots that should have feeder roots but do not. The absence of feeder roots may be a clue that either phylloxera or nematodes are at work. On older sections of the roots, look for the characteristic "charred" look of dark, cracked sections. Carefully wash the soil off of a section of root. Use a 10× hand lens in the field or, better yet, a 40× microscope in the laboratory to look for phylloxera on the sample. Commercial laboratories that conduct nematode analyses can sometimes check for phylloxera.

Control measures. There is no organic control measure available for an existing population of phylloxera. Some practitioners claim that infestations can be managed with careful water and soil management, but this has not been scientifically tested. Vigorous vines on deep, fertile soils with larger rooting areas will be able to resist an infestation longer, but ultimately, most own-rooted vines succumb. Therefore, the only secure management for phylloxera is the planting of scion stock on resistant rootstock with no *V. vinifera* parentage. Allowable rootstock parentage includes *V. riparia* (e.g., Riparia Gloire), *V. rupestris* (e.g., St. George), *V. riparia* × *V. rupestris* crosses (e.g., 3309, 101–14, Schwarzman), *V. berlandieri* × *V. riparia* crosses (e.g., SO4, 5C, 420A), and *V. berlandieri* × *V. rupestris* crosses (e.g., 110R, 1103-P). The grower's choice of rootstock also depends on other characteristics, including vigor, mineral absorption rate, effect on fruit set and maturity, and resistance to other pests (such as nematodes).

REFERENCES/RESOURCES

General

Flaherty, D. L., L. P. Christensen, W. T. Lanini, J. J. Marois, and L. T. Wilson. 1992. Grape pest management, 2nd ed. Oakland: University of California, Division of Agriculture and Natural Resources, Publication 3343.

Thrupp, L. A., M. J.Costello, and G. McGourty. 2008. Biodiversity conservation practices in California vineyards: Learning from experiences. California Sustainable Winegrowing Program Bulletin, San Francisco.

Chemicals Registered for Organic Grape Production in California

California Department of Pesticide Regulation (DPR). 2008. Product/Label Databases. Sacramento. www.cdpr.ca.gov.

Organic Materials Review Institute (OMRI). 2008. OMRI Products List. OMRI, Eugene, Oregon. www.omri.org.

IPM Theory and Practice

Flint, M. L., and P. Gouveia. 2001. IPM in practice: Principles and methods of integrated pest management. Oakland: University of California, Division of Agriculture and Natural Resources, Publication 3418.

Norris, R. F., E. P. Caswell-Chen, and M. Kogan. 2002. Concepts in integrated pest management. Lebanon, Indiana: Prentice-Hall.

Pedigo, L. P., and M. E. Rice. 2005. Entomology and pest management, 5th ed. Lebanon, Indiana: Prentice-Hall.

Organic Grape Production

Daane, K. M., R. H. Smith, K. M. Klonsky, and W. J. Bentley. 2005. Organic vineyard management in California. IPM in Organic Systems, XXII International Congress of Entomology Symposium, Brisbane, Australia, 16 August 2004.

Leafhoppers

Costello, M. J. 2008. Regulated deficit irrigation and density of *Erythroneura* spp. (Hemiptera: Cicadellidae) on grape (*Vitis vinifera*). Journal of Economic Entomology 101:1287–1294.

Costello, M. J., and K. M. Daane. 2003. Spider and leafhopper (*Erythroneura* spp.) response to vineyard ground cover. Environmental Entomology 32:1085–1098.

Daane, K. M., and L. E. Williams. 2003. Manipulating vineyard irrigation amounts to reduce insect pest damage. Ecological Applications 13…1650–1666.

Daane, K. M., G. Y. Yokota, Y. D. Rasmussen, Y. Zheng, and K. S. Hagen. 1993. Effectiveness of leafhopper control varies with lacewing release methods. California Agriculture 47(6):19–23.

Jepson, S. J., J. A. Rosenheim, and M. E. Bench. 2007. The effect of sulfur on biological control of the grape leafhopper, *Erythroneura elegantula*, by the egg parasitoid *Anagrus erythoneurae*. BioControl 52:721–732.

Lepidoptera

Mills, N., C. Pickel, S. Mansfield, S. McDougall, R. Buchner, J. Caprile, J. Edstrom, R. Elkins, J. Hasey, K. Kelley, B. Kreuger, B. Olson, and R. Stocker. 2000. Mass releases of *Trichogramma* wasps can reduce damage from codling moth. California Agriculture 54(6):22–25.

Stark, D. M., A. H. Purcell, and N. J. Mills. 1999. Natural occurrence of *Ametadoria misella* (Diptera: Tachinidae) and the granulovirus of *Harrisina brillians* (Lepidoptera: Zygaenidae) in California. Enrironmental Entomology 28:868–875.

Mealybugs

Bentley, W., L. Martin, and R. Hanna. 2001. Impact of gray field ant exclusion from vines on grape mealybug abundance, parasitism, and infestation: A progress report. UC Plant Protection Quarterly 11(3):3–6.

Breyer, L. 2006. Living with mealybugs, vine and otherwise. Practical Winery and Vineyard 28(3):43–46, 48.

Cooper, M. L., and K. M. Daane. 2007. Argentine ant management: Liquid bait program for vineyards. http://ucce.ucdavis.edu/files/filelibrary/1650/35710.pdf.

Cooper, M. L., K. M. Daane, E. H. Nelson, L. G. Varela, M. C. Battany, N. D. Tsutsui, and M. K. Rust. 2008. Liquid baits manage Argentine ants in coastal vineyards. California Agriculture 62(4):177–183.

Daane, K. M., W. J. Bentley, V. M. Walton, R. Malaka-Kuenen, J. G. Millar, C. Ingels, E. Weber, and C. Gispert. 2006. New controls investigated for vine mealybug. California Agriculture 60(1):31–38.

Daane, K. M., K. R. Sime, K. Fallon, and M. L. Cooper. 2007. Impacts of Argentine ants on mealybugs and their natural enemies in California's coastal vineyards. Ecological Entomology 32:583–596.

Smith, R., and L. Varela. 2006. Which mealybug is it, and why should you care? Practical Winery and Vineyard 27(6):37–38, 40, 42–46, 125.

Varela, L. G., and R. J. Smith. 2002. Mealybugs in California vineyards. Oakland: University of California, Division of Agriculture and Natural Resources, Publication 21612.

Welch, M. D. 2008. Interactions among weeds, ants, and obscure mealybug (Pseudococcus viburni) in central coast vineyards. M.S. Thesis. California Polytechnic State University, San Luis Obispo.

Mites

Church, E. R. G. 2009. Efficacy of botanical and mineral oils on Willamette mite (Acari: Tetranychidae). M.S. Thesis. California Polytechnic State University, San Luis Obispo.

Costello, M. J. 2007. Impact of sulfur on density of Tetranychus pacificus (Acari: Tetranychidae) and Galendromus occidentalis (Acari: Phytoseiidae) in a Central California vineyard. Experimental and Applied Acarology 42:197–208.

Hanna, R., L. T. Wilson, F. G. Zalom, and D. L. Flaherty. 1997. Effects of predation and competition on population dynamics of Tetranychus pacificus on grapevines. Journal of Applied Ecology 34:878–888.

Hanna, R., F. G. Zalom, and L. T. Wilson. 1997. 'Thompson Seedless' grapevine vigor and abundance of Pacific spider mite (Tetranychus pacificus McGregor) (Acari: Tetranychidae). Journal of Applied Entomology 121:511–516.

Macmillan, C. W. 2005. A protocol for using the mite brushing machine for measuring levels of Willamette Spider Mite (Eotetranychus willamettei McGregor) on grapes. M.S. Thesis. California Polytechnic State University, San Luis Obispo.

Phylloxera

Campbell, C. 2004. Phylloxera: How wine was saved for the world. London: HarperCollins.

Granett, J., A. Wlaker, L. Kocsis, and A. D. Omer. 2001. Biology and management of grape phylloxera. Annual Review of Entomology 46:387–412.

Sharpshooters

Biological Control of Glassy-Winged Sharpshooter in California. Applied Biological Control Research Laboratory, Department of Entomology, University of California, Riverside. www.biocontrol.ucr.edu/gwssbiocontrol.html.

10 Managing Vertebrate Pests in an Organic Vineyard

GREGORY A. GIUSTI

Vertebrates, those animals with backbones (fish, amphibians, reptiles, birds, and mammals), can be contentious pests in farming. Though they are readily recognized as having the potential to cause significant damage in agricultural systems, they also elicit strong emotional outcries from a public sensitized to the plight and pain of animals.

It is imperative, then, that a grower's vertebrate pest management effort have as its primary focus the reduction or elimination of the damage caused by an animal rather than the lethal removal of that animal. There are strategies you can consider that will help you reduce the damage caused by wildlife but do not always result in a pest animal's death.

VERTEBRATE PEST IMPACTS ON VINEYARDS

Several vertebrate species can cause economically significant damage to vineyards. These include pocket gophers (*Thomomys*), voles or meadow mice (*Microtus*), ground squirrels (*Spermophilus*), hares (*Lepus*), rabbits (*Slyvilagus*), deer (*Odecoileus*), and a number of bird species. Pocket gophers are probably the most widespread and significant pests. They cause damage by feeding on the roots and stocks of vines, gnawing on irrigation lines, digging burrows that divert irrigation water and have the potential to weaken storage dams, and building mounds that impede other management practices. Voles, ground squirrels, rabbits, and hares may also cause significant damage by feeding directly on the vines and gnawing on plastic irrigation equipment. Deer's excessive browsing can stunt vines or limit the plants' ability to produce photosynthate. Birds feed directly on the crop and have the potential to reduce yields.

MANAGEMENT STRATEGIES AFFECTING VERTEBRATES

Cover crops. Current knowledge indicates that cover cropping in vineyards can increase mammal pest problems by providing additional food and cover resources that favor higher populations of some species (Whisson and Giusti 1998). However, the magnitude of the effect may depend on the location of the vineyard, the pest species, the pest's history in the area, and the type of cover crop selected.

The potential effect of a cover crop on vertebrate pests should be considered prior to planting. If the benefits of a cover crop outweigh the added costs and energy associated with vertebrate pest control, cover cropping makes sense. Furthermore, the anticipated pest problems may be offset by the way you manage the cover crop and the plant species you select. For example, you may be able to reduce the significance of rodent impacts in a newly planted vineyard by delaying your cover crop planting by 1 or 2 years until the vines are large enough to sustain some feeding damage without ill effect.

Fencing. Constructing a fence is the most effective way to eliminate feeding by many vertebrate species (e.g., deer, rabbits, and hares). Cost is a consideration, however; a good fence is expensive to build and maintain, and fences can have a negative effect on animals' natural movement patterns through the environment. Take particular care to minimize the acreage affected by fences, and consider providing pathways or corridors that will allow for animal movements, particularly along watercourses where access to water and other habitats may be essential to wildlife.

Netting. Though the installation of netting on vines does involve a substantial initial expense, research has shown netting to be the most efficacious method of minimizing bird damage. Netting is often used in conjunction with fencing as a way to address both birds and roaming mammals in areas with heavy animal damage. The process of applying or removing netting on the vines can be tedious. Some bush berry growers have designed a system where the netting is elevated above the crop by poles and suspended on wires so they can retract it more or less the same way that you draw window curtains in a home (Figure 10.1).

Repellents. Chemical repellents can be effective against deer and rabbits. In order for a repellent to be effective, it must be applied often and the vine must be fully covered with the solution. Overhead irrigation, ambient atmospheric moisture (fog, rain, dew, etc.) can dilute the repellent and so require more frequent applications. The use of repellents in organic vineyard pest management is constrained by their terms of registration and by the requirement that they be approved by the appropriate organic certifier.

Traps. Trapping can be highly effective if the person responsible for the traps is diligent and thorough. The most widely used trap styles include the Macabee and Cinch traps. Instruction sheets for their proper use and placement are available where you buy the traps, or in some cases online at the manufacturer's website. Trapping is the most effective choice for gopher control available to organic farmers. There are also traps designed to capture birds, the most commonly used design being the modified Australian crow trap, which can be used to trap house sparrows, house finches (linnets), and starlings (Figure 10.2).

Dispersal techniques. Propane canons (Zon Guns, firecrackers, or shell crackers [Figure 10.3]) can be effective against birds if used in combination with other devices (e.g., Mylar tape, bird bombs, kites, etc.). To improve their efficacy, you should relocate sonic devices regularly so the birds will not become acclimatized. Because of the extreme noise levels generated by some sonic devices, you need to take into account their potential effect on adjoining properties and nearby neighborhoods when considering their use. New research on the use of some species' own recorded distress calls against those species (e.g., common crows) shows promise for that practice as a means of dispersing birds from an area.

Ultrasonic devices. Commercially made ultrasonic devices are available for use as vertebrate pest repellents but they have not proven to be effective.

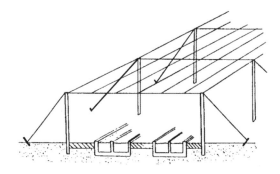

Figure 10.1. One arrangement for wires extended above a crop to suspend netting materials. *Illustration* from *Prevention and Control of Wildlife Damage* (Hygnstrom, Timm, and Larson 1994).

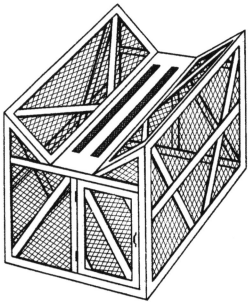

Figure 10.2. Photograph and diagram of an Australian crow trap design modified for crowned sparrow control. Assembled trap measures 6 feet wide, 8 feet long, and 6 feet tall. For construction details, see *Prevention and Control of Wildlife Damage* (Hygnstrom, Timm, and Larson 1994). *Photo* by Jack Kelly Clark.

PEST-SPECIFIC STRATEGIES

Pocket Gophers

Food quality and quantity are the major factors influencing the distribution and abundance of pocket gophers (Figure 10.4). Because their breeding is regulated by the availability of green forage, pocket gophers may breed year-round under irrigation conditions (e.g., in alfalfa fields or vineyards). Planting certain types of cover crop can actually increase the available food supply, favoring this pest and potentially increasing its numbers.

Although the feeding habits of pocket gophers suggest that they prefer cover crops over vines, they may be forced to feed on the vines when the cover crop dies or is removed. In this case, pocket gopher numbers may be unusually high as a result of cover cropping, and the ensuing damage to vines can be quite severe.

Site preparation. Disking and tilling operations can reduce the number of pocket gopher tunnels in a planting site since most burrows are between 8 and 12 inches (20 and 30 cm) below the surface, shallow enough to be disrupted by standard tractor-drawn implements. No direct effect on gopher populations as a result of disking has been demonstrated, but the effectiveness of limiting a species' ability to occupy a site or become established has been proven for other rodent species. After disking, any new burrowing activity should be readily visible, making trap placement and other control techniques fairly straightforward.

If possible, try to control pocket gophers in neighboring areas that adjoin the vineyard as well to reduce the risk of subsequent invasions from those areas.

Cover crop selection. Pocket gophers are strict herbivores. They exhibit a preference for fleshy and succulent roots and stems over grasses, which have a fibrous root system. Food preferences appear to correspond with the fiber-to-protein ratio of the plant. Gophers are able to assimilate more protein from plants that are low in fiber.

Pocket gophers are adaptable in their feeding habits and will find alternate food sources when their preferred foods are scarce, but their population usually declines under such circumstances. For this reason you will want to consider pocket gophers' food preferences when you select a cover crop blend. Cover crops that include grasses or cereals with

Figure 10.3. Shell crackers are fired from a 12-gauge shotgun. They produce an aerial explosion and can be useful in frightening birds out of fields or away from roosts. *Illustration* by Jill Sack Johnson, from *Prevention and Control of Wildlife Damage* (Hygnstrom, Timm, and Larson 1994).

Figure 10.4. Pocket gopher. *Photo* by Jack Kelly Clark.

fibrous root systems rather than strongly tap-rooted legumes may limit the amount of food available to gophers and thereby reduce their reproductive potential. Some grasses (e.g., California brome) are high in moisture content and so should be avoided.

Trapping. Assuming that the person responsible is serious and diligent about the assignment, trapping can be a very effective option. Trapping, like any pest management practice, requires frequent, regular monitoring and action. A single pocket gopher in a young vineyard can damage several vines, resulting in economic losses for as many years as it takes for the vines to be replaced and reach the ruined vines' former potential. Trapping programs provide both a direct control strategy and a way to monitor the number of animals removed from a given area. Traps give an organic grower the potential to dramatically reduce population numbers while avoiding any chance of secondary poisoning (Figure 10.5).

Voles *(Microtus)*

Because it provides voles (meadow mice) (Figure 10.6) with food, conceals them from predators, and protects them from unfavorable weather, ground vegetation is the most important factor affecting vole abundance. A vineyard with cover crops most certainly favors vole populations. Without careful monitoring, vole damage may be severe enough to kill several vines, especially in a young planting.

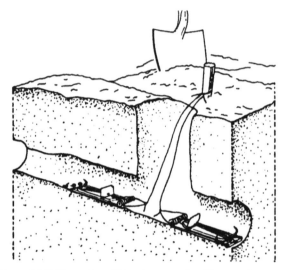

Figure 10.5. Illustration of Macabee traps faced in opposite directions in a pocket gopher burrow.

Figure 10.6. California vole. *Photo* by Jack Kelly Clark.

Vole populations can be expected to cycle throughout the year. Usually, the first peak will come after the first rains of autumn. A second, more prominent spike usually occurs in March and April. If the vineyard is planted with cover crops, they can easily conceal the visible signs of increased vole numbers.

Site preparation. Organic approaches to minimizing vole damage should focus on vegetation management. If you are using cover crops, strive to reduce or eliminate plants that come in contact with vines. Maintain a clean strip of soil directly beneath the vines to eliminate any direct contact or cover for the mice, and they will be less likely to girdle vines. Practices that reduce suitable cover in surrounding areas also play a role in preventing serious vole problems.

Vine protection (exclusion). You can guard young vines from girdling by protecting them with cylinders made from hardware cloth, sheet metal, or heavy plastic placed around the trunk. Plant tubes, milk cartons, or any other devices that place a physical barrier between the vine and voles can effectively deter feeding damage. Place the tubes or cartons so they are in contact with the soil all the way around. If you use paper cartons, check them after 1 year to ensure that they still provide a secure barrier against feeding.

Cover reduction. Frequent close mowing of ground cover is a beneficial practice for vineyards where it is possible. It removes protective cover for voles and makes it easier for predators to find them. Do not leave the clippings along the vine rows after mowing, since voles may find shelter under the clippings and continue to damage your vines.

Rabbits and Hares

Rabbits and hares are known to frequent vineyards and forage on whatever ground cover is available. Cover crops can serve as a powerful attractant, so take care when planting alfalfa, clovers, vetches, beans, or peas in an area with a high rabbit population. Fortunately, grapevines grow beyond the reach of most rabbits and hares after just 1 year. Newly planted vineyards, however, are at risk from feeding damage to the canes. Take care to protect newly planted vines with exclusionary tubes or fences. Rabbit fences should have a mesh size no larger than 1 by 1 inch (2.54 cm by 2.54 cm). If you are using a fence to keep out both rabbits and deer, the smaller mesh should extend at least 3 feet (1 m) above the ground. Once the vines are established, rabbits and hares may continue to feed on the ground cover in the vineyard, but the vines themselves should be well out of reach.

Ground Squirrels

Ground squirrels tend to disappear from land that is under complete and frequent cultivation (Figure 10.7). They will, however, maintain burrow systems along fence lines, road right-of-ways, and other uncultivated areas, and they can travel 100 yards (90 m) or more from their burrow to feed in adjacent crops.

At first when they emerge from hibernation in spring, ground squirrels feed almost exclusively on green vegetation. When annual grasses and forbs begin to produce seed and dry up, the squirrels start to eat seeds and fruits as well as bark from vines. The presence of cover crops in a vineyard at this time may encourage squirrels to spend more time in the vineyard, giving them the potential to cause more damage to vines.

Lethal organically approved control methods include incendiary sulfur-based smoke bombs, traps, and shooting. Since standards and rules governing the use of certain lethal vertebrate control measures are subject to change, you should always check with your organic certifying inspector to ensure that incendiary devices are still an approved option under your current organic program.

Incendiary devices emit dense smoke to displace oxygen in ground squirrel burrows and are placed directly into the burrows after the animals have emerged from hibernation. It is important to follow all label directions, taking into account soil moisture conditions and proper application techniques. Both factors can have a significant effect on the efficacy of the devices.

Trapping. The most commonly used lethal traps for ground squirrels are the California box trap and the Conibear 110 trap. The Conibear trap is a nonbaited, spring-loaded device that is placed directly over the burrow entrance after the animals have emerged from hibernation. The box trap (Figure 10.8) can be used in conjunction with an attractive bait (rodent chow or hen scratch) and usually is placed in the aboveground runways that connect burrow openings and foraging areas.

Other nonlethal control methods include the removal of any piles of prunings and other debris from vineyards and their margins. Vineyard margins should be kept mowed.

Deer

Deer are large herbivores and can cause significant damage to vineyards. They are resident in most winegrape growing regions of the state and spend the entire year in the vicinity of the place where they were born. Deer are game mammals, so there are no lethal options for minimizing deer damage outside of regularly scheduled hunting seasons and bag limits. However, other options are available to reducing deer impacts on vineyards.

Figure 10.7. Ground squirrel. *Photo* by Jerry P. Clark.

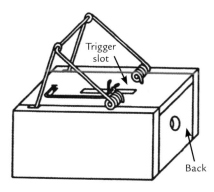

Figure 10.8. Box traps should be placed on the ground near squirrel burrows or runways.

Fencing. Both electric and traditional fencing can be very effective at keeping deer feeding damage to a minimum. To keep costs low and maintain wildlife habitat, the fencing should only surround the planted area, allowing wildlife to use other portions of a grower's property. Electric fences provide for a temporary, relatively easy-to-move solution if you have deer that are affecting young plantings. Any permanent fencing should be at least 8 feet (2.5 m) high. If you have large vineyard blocks, include wildlife corridors in the vineyard design so deer and other wildlife can pass through your property.

The California Department of Fish and Game (DFG) expects landowners to use fencing where deer might enter vineyards. Landowners cannot expect DFG to issue a depredation permit allowing them to shoot wildlife whenever it becomes a problem in an unfenced vineyard.

Repellents. Most deer repellents are not registered for use in winegrape vineyards. Inherent constraints on the manner and timing of their application make them difficult to use. Repellents generally are not recommended for use in commercial plantings.

Birds

Because of their protected status and their ease of mobility, birds pose the greatest challenge to vineyard managers experiencing crop damage. Starlings (Figure 10.9), house sparrows, and rock doves (common pigeons) may be lethally controlled without special permit. Most blackbirds, crows, and magpies can be lethally controlled if they are damaging or threatening vineyards. House finches can be controlled under

Figure 10.9. European starlings can be destructive pests in maturing grapes late in the growing season. *Photo* by Jerry P. Clark.

the general supervision of a local county agricultural commissioner, so growers should check with the commissioner's office before undertaking any control measures. To kill most other bird species, you need a permit issued either by the California Department of Fish and Game or the Federal Fish and Wildlife Service for migratory species (ravens, waterfowl, doves, band-tail pigeons, etc.).

Netting. This provides the most efficient means of minimizing bird feeding damage to grapes. Newly developed application machinery and techniques make the placement and removal of nets over vines a viable option today. Though expensive on initial installation, the newest net materials have UV light protection that helps them to survive for up to 10 years. Experts recommend that the nets be placed over the vines at veraison and tied shut to keep birds from entering from the ground. If grape losses from birds exceed 5 percent, netting is the best option for bird control.

Sonic devices. Most traditional sonic devices and pyrotechnic devices are based on the notion that a loud, frightening noise will cause birds to move to another location. Birds are sapient creatures, though, so they can learn from their surroundings. When exposed to a repetitious sound from a stationary source, regardless of the sound's intensity, most bird species will eventually become acclimated to it and come to regard the sound as part of the local background noise. There is anecdotal evidence to suggest that some birds are actually attracted to a vineyard equipped with a sonic device, since the birds learn that the sound is associated with ripening fruit.

Traditional sonic devices such as Zon Guns (a brand of propane cannon) may have some utility, but most scientific literature supports their use only in combination with other frightening tactics such as kites or Mylar tape. When used, this type of sonic device should be moved frequently within the vineyard to keep the birds from becoming too comfortable with its location. With continued urbanization of California's agricultural lands, these devices will become more and more problematic, since the loud noise will certainly annoy neighboring residents. You have to take adjoining properties and neighbors' interests into account when planning your bird-control strategies.

A newly emerging branch of science, biosonics, strongly suggests that some bird species (including crows and possibly starlings) are sensitive to recordings of their own species' distress calls played over loudspeakers. The birds do not appear to become acclimatized to this sound, giving it a greater potential for long-term use. Unfortunately, house finches (linnets) have not shown the same level of response to recorded distress calls.

Repellents and reflectors. Though a number of repellent materials have emerged recently to help growers minimize bird feeding damage to grapes—including Mylar tape, kites, and scare-eye balloons—none has been systematically tested for effectiveness. Furthermore, though each of these frightening devices may have some degree of efficacy, you must keep in mind that they are intended to reduce damage, not fully eliminate it over large areas.

The keys to an effective bird dispersal program are timing, persistence, organization, and diversity of tactics. Properly used, frightening devices can be an effective tool for dealing with potential health and safety hazards, depredation, and other nuisances caused by birds.

PEST MANAGEMENT GUIDELINES

In every pest management situation, growers must be diligent and watch for telltale signs of problems. A regular monitoring program for the early detection of increases in pest populations and damage is important. In addition to monitoring the vineyard, the grower also needs to keep a routine watch on surrounding areas, watching for signs of animals that may have the potential to re-infest the vineyard. Inspect cropped and adjacent areas for species such as pocket gophers, and watch daily for birds when the crop is accumulating sugars. The information contained in Table 10.1 may prove helpful as you develop your monitoring program and assess management options.

REFERENCE

Whisson, D. A., and G. A. Giusti. 1998. Vertebrate pests. Pp. 126–130 in C. Ingels (ed.), Cover cropping in Vineyards. University of California, Division of Agriculture and Natural Resources, Oakland. Publication 3338.

Table 10.1. Vertebrate pest species, signs of damage, and control considerations

Species	Signs	Alternative controls and options
Deer	stripping of foliage from vines, breakage or scarring of young vines, presence of scat and footprints	cover crop selection (grasses rather than legumes), exclusion fences
Ground squirrel	girdling of vines aboveground, feeding on foliage and fruit, gnawing on irrigation lines, observation of daytime activity, burrow systems (especially on perimeter of vineyard)	incendiary sulfur bomb burrow fumigants, shooting, debris pile and pruning removal
Pocket gopher	plugged burrow systems, earth mounds, girdling of vines belowground, stunted vines, damage to irrigation lines	cover crop species selection (grasses rather than legumes), mowing to reduce and facilitate early detection, flood irrigation, trapping
Rabbits and hares	feeding on foliage and fruit, girdling or complete cutting of vines aboveground, observation of activity	exclusive fencing, vine guards, shooting, delay cover crop plantings for 1 or 2 years in new vineyard
Vole	runways and open burrow entrances, presence of scat, girdling of vines aboveground	cover crop species selection (erect bunch-type growth or short plants, avoid high-moisture plants), maintenance of cover-free strip around base of vines, mowing to reduce cover, vine guards, trapping

Organic Vineyards and the Environment

PHOTO: TOM LIDEN

Biodiversity, Habitat, and Natural Resource Issues in Winegrape Production

ROBERT L. BUGG, GREGORY A. GIUSTI, ADINA MERENLENDER, SUSAN P. HARRISON, GLENN T. McGOURTY, AND KENDRA BAUMGARTNER

> . . . [C]omposition, structure, and function . . . determine, and in fact constitute, the biodiversity of an area. Composition has to do with the identity and variety of elements in a collection, and includes species lists and measures of species diversity and genetic diversity. Structure is the physical organization or pattern of a system, from habitat complexity as measured within communities to the pattern of patches and other elements at a landscape scale. Function involves ecological and evolutionary processes, including gene flow, disturbances, and nutrient cycling.
>
> — Noss (1990), in part paraphrasing Franklin (1988).

Biodiversity, literally "the variety of life," includes in its scope organisms and their relationships to each other and to the broader environment. Biodiversity can be regarded as a key natural resource, and a practical understanding of it can aid us in preserving other natural resources. Although an individual may own title to a plot of land, the public resources that occur on that land are held in the public trust for the benefit of all (see Haddad 2003). All people share responsibility for protecting and maintaining natural resources, including air, water, native plants, fish, and wildlife. As noted by Moonen and Barberi (2008), the incentives for preserving biodiversity in agroecosystems include:

1. Species, community, habitat or overall biodiversity conservation regardless of its functions

2. Biodiversity conservation to attain production and environmental protection services

3. Use of bio-indicators for agroecosystem monitoring

As stated by the USDA National Organic Standards Board, "Organic farming is an ecological production management system that promotes and enhances biodiversity, biological cycles, and biological soil activity." The contributions of vineyards to biodiversity conservation and restoration must ultimately be viewed from a landscape- to regional-scale perspective, in context with other land uses (Rustigian et al. 2003; Santelmann et al. 2004; Hilty et al. 2006; Zanini et al. 2008).

Such a comprehensive view, though, is beyond the scope of this chapter. The degree of biodiversity in and around vineyards can be changed either by adding or removing species or by managing existing species in ways that change their spatial and temporal relationships and interactions.

In California, many vineyardists have opted for the "clean look," removing all vegetation from the area surrounding the vineyard, usually by means of herbicides or tillage. While some might like the way this looks, the lack of living plant cover other than that provided by grapevines may reduce the abundance of spiders and other natural enemies of insect and mite pests, as well as other desirable wildlife. Moreover, vegetation cover, even that provided by weeds, may, through its addition of biodiversity composition, structure, and function, reduce erosion and improve water quality as compared to bare ground.

Sidebar 11-A

BIODIVERSITY

Biodiversity (short for "biological diversity") is a term that includes the variety of life in all forms, levels, and combinations and includes ecosystem, species, and genetic dimensions. To grasp the concept of biodiversity, you need to recognize not just the species composition, but also the structure and function of an ecosystem. When you change the plants or animals within a biodiverse community you affect ecosystem processes and the natural goods and services that we rely on, such as clean water, fertile soil, and healthy fish populations.

Biodiversity is manifest on a number of levels in an environment:

- *Landscape and physical environment level:* A large surrounding area made up of a mosaic of different habitats, such as lakes, forests, and grasslands.

- *Ecosystem level:* Natural communities and their environment functioning as an ecological unit.

- *Species and population level:* Plant and animal elements that share resources (i.e., food, water, space) within an ecosystem.

- *Genetic composition level:* The inherited gene sequences that determine various characteristics of an individual within the species.

These are the basic elements of a biodiverse environment:

- **Species composition:** The set of organisms that populate an environment.
  - **Landscape:** The species composition of a landscape includes all the existing, living, or formerly living landscape elements, which can be there as a result of natural or artificial processes.
  - **Ecosystems:** The species in natural communities that, with their environment, function as an ecological unit.
  - **Species/populations:** The living species (plant and animal) dwelling in an ecosystem.
  - **Genetic composition:** Hereditary factors in each individual of a species, based in

DNA sequences and subject to change over generations through the forces of evolution (e.g., mutation, recombination, natural selection).

- **Structure:** The physical organization of an environment, made up of interrelated parts.
  - **Landscape:** The structural configuration of landscape and geographical features.
  - **Ecosystems/habitats:** The type of physical structures created by the adaptation and specialization of vegetation to other habitat elements such as rocks and cliffs.
  - **Species/populations:** The relative level of isolation, resulting from aspects of an environment's structure, between different populations of plants or animals of a given species.
  - **Genetic structure:** The relative isolation, resulting from aspects of an environment's structure, of genes between individuals within a same-species population.

- **Function:** Interactions among biota and the physical environment that result in operations or processes, such as nutrient cycling.
  - **Landscape:** Natural phenomena that shape the larger landscape (e.g., earthquakes, uplift, erosion and deposition by rivers and streams, glaciations).
  - **Ecosystem:** The functions of energy and nutrients that move through the landscape's various ecological systems. For example, water flows, energy flows (sunlight -> plant photosynthesis -> herbivore ingestion -> carnivore ingestion), nutrient flows (photosynthate -> leaf growth -> detritus from fallen leaves -> decomposition into basic elements of N, P, K and soil constituents).
  - **Species/populations:** Demographic processes, life histories.
  - **Genetic processes:** At the local level, gene flow between individuals within a community or population. At a landscape level, gene flow between separate populations.

Sidebar 11-B

BIODIVERSITY WORKSHEET *(Adapted from Giusti et al. 1996)*

There are many ways for winegrowers to manage their vineyards to maintain and improve biological diversity. We recommend a systemwide approach to biodiversity management that includes a recognition of the relationship that exists between the composition, structure, and function of the various hierarchical levels discussed earlier (see "Biodiversity" sidebar). Many management decisions and actions can have unforeseen consequences.

This worksheet is designed to guide you through a management decision matrix that will help you determine whether unintended consequences are having a negative impact on biodiversity at the landscape, community, or population level.

Composition (fill out those that apply)	Select location*	Habitat elements present (check all that apply)	Size and scope (number of trees or acres)
Landscape type: ☐ Interior valley ☐ Interior mountains ☐ Coastal valley ☐ Coastal mountains ☐ Other Total acreage of the enterprise area: _____acres		☐ Live trees ☐ Conifers ☐ Hardwoods ☐ Dead trees ☐ Rock piles ☐ Tules/cattails ☐ Willows ☐ Cover crops ☐ Ponds ☐ Floating vegetation ☐ Streams ☐ Riparian vegetation	

*Use a separate copy of this worksheet for each landscape type.

If you checked any of the habitat elements above:

- Can you think of ways to minimize the impact of any of these elements through your management techniques?
  - Are there any adaptations you can make to your management techniques to minimize impacts?
  - Are there design modifications you could make to minimize impacts to these elements?
- Are there ways to expand the current size or distribution of the elements you have identified?
  - Are there areas within the ranch that you can allow to rest so that some of these elements could naturally expand, reducing the cost and time commitments required of you?
- Are there ways to reduce systemic hosts of Pierce's disease along riparian zones, replacing that vegetation with more desirable native plants?

If you left any of the habitat elements unchecked:

- Can you identify areas within your ranch where you could recruit some or all of these elements with no negative impact to your farming operation?
- Could any of these elements be readily recruited if you altered or eliminated a current practice (e.g., mowing or spraying roadside ditches that kill tules, cattails, or valley oaks)?
- Are there areas of your ranch that you could allow to "go natural" with no negative impact to your farming operation?

GETTING HELP

Financial incentives are available to encourage landowners to plant hedgerows. Here are some programs to consider:

- **Environmental Quality Incentives Program** (USDA Natural Resources Conservation Service). Cost sharing is available for various conservation practices on eligible land. http://www.nrcs.usda.gov/programs/eqip/
- **Wildlife Habitat Incentives Program** (WHIP) (USDA Natural Resources Conservation Service). A voluntary program for landowners who want to develop and improve wildlife habitat on private lands. http://www.nrcs.usda.gov/programs/whip/
- **Conservation Reserve Program** (USDA Farm Services Agency). The CRP provides landowners with financial assistance to convert environmentally sensitive areas to vegetative cover. http://www.nrcs.usda.gov/programs/crp/
- **Partners for Fish and Wildlife Program** (U.S. Fish and Wildlife Service). This program provides landowners with financial assistance to help them voluntarily restore wetlands or other critical habitat for fish and wildlife on their property. http://partners.fws.gov/

In some cases, enhancing biodiversity can enable a grower to reduce reliance on agricultural chemicals, as when a grower uses biological control agents or nitrogen-fixing cover crops as partial substitutes for synthetic pesticides and fertilizers; these kinds of functions are termed "agroecosystem services" (Moonen and Barberi 2008). Biodiversity can be enhanced to improve aesthetics, manage nutrients, control erosion, enhance crop yield and quality, and support wildlife. Practical management of biodiversity is especially important in organic agriculture because organic growers have no recourse to synthetic nitrogen fertilizers and pesticide applications.

There are many ways for a vineyard manager to maintain or enhance biodiversity and develop a more ecologically functional property, enabling the grower to stay in compliance with environmental regulations, improve fish and wildlife habitat, and generally practice responsible land stewardship. This approach is also usually viewed favorably by neighbors and other interest groups and helps to ease tensions among competing interests.

BIODIVERSITY IN CALIFORNIA AND THE IMPACT OF "VINEYARDIZATION" OF LANDSCAPES

California is one of the world's biodiversity hot spots, with a great number of native animal and plant species. The state is home to 4,400 native plant species, 48 percent of which are found in California alone. More than 20 percent of the 4,400 species are rare or endangered. California's great diversity of native species is related in part to the wide range of microclimatic and soil conditions that occur in the state. Some groups of plants remained viable in California when climatic changes devastated the same plants in other parts of North America. Mountainous areas such as the Coast Ranges of California are especially important in conserving such remnant or relict species because those areas include such a wide range of physical environmental conditions (aspect, microclimate, water, soil type) together in close proximity. This permits plant populations to adapt to climatic change by colonizing new, nearby niches as their existing habitats become unsuitable.

Native species are threatened by the introduction of nonnative competitors, agricultural and other development of wildlands, and changes in habitat caused by other human influences. Disturbance by humans also promotes invasion by nonnative competitors, many of which are adapted to continuing disturbance and have been introduced without their natural antagonists, conferring a competitive advantage relative to native organisms (Torchin et al. 2003). At first glance, it might appear that introduced (nonnative) organisms add to California's biodiversity by augmenting the state's species richness. However, these invasions often result in disease outbreaks (e.g., Pierce's disease carried by glassy-winged sharpshooter), and some introduced competitors may displace or reduce the occurrence of native species and affect ecosystem function. It is critical that we maintain and enhance the resilience of California's landscapes to future invasions through conservation of natural communities and diverse agroecosystems because monocultures are more susceptible to invasions. For a nuanced discussion of changes in species diversity at different spatial scales as a result of habitat destruction and biological invasions, see Sax and Gaines (2003).

About 1,045 nonnative plant species have become naturalized in California. Many of these are from the Mediterranean region, which has a similar climate to much of California and a much greater native plant diversity and is therefore a major potential source of invasive plants for California. About 10 percent of the naturalized exotic plants in California are regarded as threats to native biodiversity; a well-known example is yellow starthistle (*Centaura solstitialis*), which infests rangelands. In some cases, the spread of nonnative species is enabled by disturbances caused by humans through changed fire, water, grazing, and nutrient regimes. As mentioned above, the colonization and naturalization by exotic plant species could be perceived as an increase in biodiversity, but this is often not the case since some exotic species can outcompete entire native plant communities, resulting in a loss or reduction of native plants and co-evolved animals.

Agricultural and other development may lead to the loss or fragmentation of native habitats, a loss that is widely recognized as the primary cause of species and population declines here in California. Fragmentation of native habitats, defined as the habitat's loss of unity and cohesion and its breakup into isolated elements, has dramatically altered the California landscape since the Gold Rush of the mid-1800s. In the worst cases, fragmentation has resulted in the isolation of patches or islands of native habitats, interspersed among developed areas. This can disrupt animal movements, predator-prey relationships, plant pollination, and surface- and groundwater dynamics. Once the structure and composition of an area's vegetation are fragmented, fences, roads, water management, and other features and practices may further impact natural habitat and wildlife conservation.

The alteration of structure and composition of North Coast rainforests by aggressive logging, the conversion of wetlands to housing developments, and the clearing of oak woodlands, chaparral, and grasslands for vineyard establishment all contribute to habitat loss and fragmentation. As a result, California leads the nation in the number of plant and animal species listed on the Federal Endangered Species List.

Winegrapes are the only major crop in California that is expanding into wildlands, especially in the North Coast and Central Coast areas. The winegrape industry grew rapidly from the mid-1990s to 2000 as new vineyards were carved out of oak woodlands, rangelands, and farmlands that had previously been devoted to forage, field crops, and orchards. In 2000, statewide winegrape acreage stood at over 400,000 acres, up from 300,000 acres in 1995. Areas developed for winegrowing now include large sections of the Sacramento and San Joaquin Valleys, the North and Central Coast Ranges, including parts of San Luis Obispo and Santa Barbara Counties, and smaller acreages in the foothills of the central Sierra Nevada (e.g., Amador County).

Hillside vineyard development also continued in the traditional premium production areas of Mendocino, Lake, and Sonoma Counties (North Coast). Some vineyards have come into production even in nontraditional grape growing regions such as Trinity and Humboldt Counties. This rapid expansion has caused deforestation, particularly of highly diverse oak woodlands, which provide important habitat for many native plants and animals. Vineyard development poses another threat to biodiversity when winegrowers destroy native riparian plants adjacent to stream corridors in an effort to control Pierce's disease. Riparian areas in oak woodlands are important habitats for resident and migratory bird species (Giusti et al. 2003). In addition, vineyard irrigation, including that used for frost protection, leads to competition for scarce water resources and may thereby imperil native biota (Merenlender 2008; Merenlender, Deitch, and Feirer 2008). Environmental interest groups have taken issue with these and other outcomes of vineyard development and management.

As suggested above, vineyard establishment and other forms of development are putting several California vegetational complexes at risk. These include grassland, chaparral, upland woodland, and riparian woodland communities. The exact composition of

Table 11.1. Partial lists of native plant species commonly found in representative plant communities in Napa County

Grassland: Annual blue lupine *(Lupinus bicolor)* and rusty popcorn flower *(Plagiobothrys nothofulvus)*.
Mixed chaparral: Common buckbrush *(Ceanothus cuneatus)*, toyon *(Heteromeles arbutifolia)*, chaparral pea *(Pickeringia montana)*, scrub oak *(Quercus berberidifolia)*, scrub form of interior live oak *(Quercus wislizeni* var. *frutescens)*, chamise *(Adenostoma fasciculatum)*, yerba santa *(Eriodictyon californicum)*, common rush-rose *(Helianthemum scoparium* var. *vulgare)*, pitcher sage *(Lepechinia calycina)*, and deerweed *(Lotus scoparius)*.
Oak woodland: Blue oak *(Quercus douglasii)*, black oak *(Quercus kelloggii)*, canyon live oak *(Quercus chrysolepis)*, leather oak *(Quercus durata)*, scrub oak *(Quercus berberidifolia)*, interior live oak *(Quercus wislizeni)*, serrated onion *(Allium serratum)*, four-spotted godetia *(Clarkia purpurea* ssp. *quadrivulnera)*, bicolored linanthus *(Linanthus bicolor)*, downy navarretia *(Navarretia pubescens)*, and buttercup *(Ranunculus californicus)*.
Riparian woodland: Big-leaf maple *(Acer macrophylla)*, black walnut *(Juglans hindsii)*, box elder *(Acer negundo)*, California blackberry *(Rubus ursinus)*, coast live oak *(Quercus agrifolia)*, valley oak *(Quercus lobata)*, Fremont's cottonwood *(Populus fremontii)*, Oregon ash *(Fraxinus latifolia)*, poison-oak *(Toxicodendron diversilobum)*, California buckeye *(Aesculus californica)*, redbud *(Cercis occidentalis)*, red willow *(Salix laevigata)*, squaw bush *(Rhus trilobata)*, California wild rose *(Rosa californica)*, dogbane *(Apocynum cannabinum)*, mugwort *(Artemisia douglasiana)*, false lupine *(Thermopsis macrophylla)*, snowberry *(Symphoricarpos* spp.*)*, and California pipevine *(Aristolochia californica)*.

these native complexes varies depending on location. Partial lists of species common in Napa County areas are presented in Table 11.1. Many of these plant species sustain populations of insect species that are evolutionarily adapted to feed on them or pollinate them, and many of the plants provide important food for other wildlife. In addition, many insectivorous birds nest in riparian woodlands. It is important to recognize that the development of a vineyard amid wildlands necessitates the removal of native plants and other important habitat features (Figure 11.1) and thereby imperils native animals that cannot live by winegrapes alone. Also, when vineyards replace other farming systems, they may provide less-favorable conditions than other crops for foraging by some native birds, as has been shown in northern California for Swainson's hawks (*Buteo swainsoni*), which preferentially forage in alfalfa fields and in grasslands (Swolgaard, Reeves, and Bell 2008).

INCREASING BIODIVERSITY IN AND AROUND VINEYARDS

Few quantitative data are available on the occurrence of wildlife in California vineyards (but see Hilty 2001; Hilty and Merenlender 2004; Luther et al. 2008), nor do we know how wildlife diversity and density are influenced by cover cropping regimes, compost additions, or other viticultural practices. Grapevines themselves provide some resources for native wildlife. For example, informal observations in Davis, Yolo County, by Robert L. Bugg and David E. Chaney (unpublished observations, 2003) indicated that at least the following birds feed on unharvested

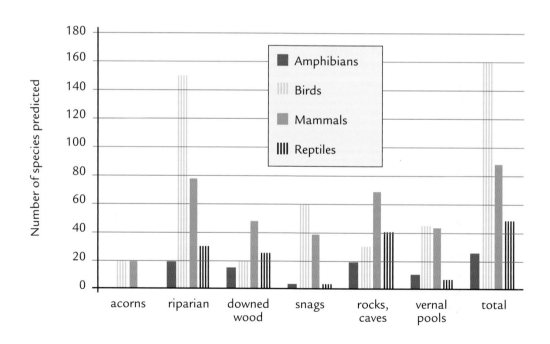

Figure 11.1. Vertebrates predicted to use habitat elements found in oak woodlands (reproduced from Giusti, Garrison, and Fitzhugh 1996).

("second-crop") grapes during late fall and early winter: American robin (*Turdus migratorius*), cedar waxwing (*Bombycilla cedrorum*), northern flicker (*Colaptes auratus*), northern mockingbird (*Mimus polyglottos*), ruby-crowned kinglet (*Regulus calendula*), western bluebird (*Sialia mexicana*), western scrub-jay (*Aphelocoma californica*), yellow-billed magpie (*Pica nuttalli*), and yellow-rumped warbler (*Dendroica coronata*). Rotting second-crop grapes sustain reproduction by vinegar flies and other Diptera, which in turn serve as prey to overwintering Anna's hummingbirds (*Calypte anna*) and perhaps other birds that eat insects. Luther et al. (2008) found that the American goldfinch (*Caruelis tristis*), chipping sparrow (*Spizella passerina*), and western meadowlark (*Sturnella neglecta*), all grassland species, were found in Sonoma County vineyards but not in other nearby landscapes (e.g., riparian zones, oak savannahs). Of course, some birds, both native and introduced, feed on winegrapes and can be pests of vineyards, including American robin and house finch (*Carpodacus mexicanus*) (Berge et al. 2007), wild turkey (*Meleagris gallopavo*), California quail (*Lophortyx californica*), European starling (*Sturnus vulgaris*), and crown sparrows (*Zonotrichia* spp.) (Giusti, personal observation).

The simple choice to allow volunteer native trees to persist along roadside ditches or fence lines will eventually lead to improved structural biodiversity. The large woody structure that eventually results will provide perches for raptors seeking rodent prey; these perches also support bird species that feed on the grapes. By contrast, by permitting tules (*Scirpus* spp.) or cattails (*Typha* spp.) to grow in ditches, you provide habitat that supports nesting marsh wren (*Cistothorus palustris*), a "neutral" species (it does not feed on winegrapes, but provides no actual known benefit to vineyards).

An easy way to maintain biological diversity is to retain, where possible, the structure and composition of vegetation that existed on the site prior to the vineyard project. Retained trees, bushes, hedgerows, riparian vegetation, rock piles, logs, and snags (standing dead trees) will provide shelter, moisture,

Sidebar 11-C

A SIMPLE ACT CAN HAVE A BIG IMPACT.

With planning and forethought, people can improve biological diversity on their property while managing a winegrape crop. A 2003 survey asked growers what practices they had incorporated into their management regime to increase biodiversity and received these responses (the numbers in parentheses indicate how many of the 22 growers surveyed had incorporated the practice):

- Certified organic or using organic methods (4)
- Conservation easements (1)
- Contour planting (1)
- Controlled grazing (2)
- Cover cropping (11)
- Endangered species management (3)
- Fisheries management and restoration (5)
- Habitat buffers (6)
- Integrated pest management (12)
- Invasive species management (3)
- Low-input farming/no-till practices (7)
- Mulch management (3)
- Native plant restoration (8)
- Oak planting and protection (7)
- Owl/raptor/bat boxes and perches (10)
- Participation in watershed management group (3)
- Ponds managed for wildlife (5)
- Riparian/wetland restoration and enhancement (9)
- Runoff management/water quality management (9)
- Water conservation (4)
- Wildlife corridors/wildlife-friendly fencing (10)
- Wood duck boxes (5)

and food for animals. By identifying native plant communities and conserving them, you will ensure continued gene flow between individual plants and support dependent insect populations.

There are also more formal measures that may somewhat mitigate the net loss of biodiversity that occurs through the "vineyardization" of landscapes. We will address several of these in detail:

1. Water-related issues, including irrigation practices and management of reservoirs, riparian zones, and other water-related features to favor native biota (Kolozsvary and Swihart 1999; Hazell et al. 2001; Joly et al. 2001; Maezono and Miyashita 2003; Hazell et al. 2004; Davies et al. 2008a; Davies et al. 2008b; Ruggiero et al. 2008; Zanini et al. 2008)

2. Use of "resident vegetation" or seeded cover crops in alleys (also called "middles")

3. Designing and implementing wildlife-friendly fencing

4. Hedgerows (Earnshaw 2004)

We will only mention in brief here the use of bird and bat boxes to partially substitute for the cavities once provided by native trees; this subject is addressed in an excellent discussion in Heaton et al. (2008).

Water-Related Issues. Water is essential to all life, but conflicts arise between humans' demand for water (and their concomitant need to control and contain it) and the sometimes competing need for water in natural ecosystems to support wild plants and animals.

On floodplains, the levees that border streams and rivers are essential tools for flood management in many areas, yet they have unintended consequences for biodiversity when they limit the natural tendency of a river to meander. Levee confinement leads to loss of natural river features such as sloughs, oxbow lakes, and lateral pools, which are important in the biology of many plants and animals, especially some species of fish. It also directs the river's erosive force downward, leading to deeply incised channels. This is especially true when levees are used in stream reaches below dams, because streams in such reaches often have reduced sediment load and greater erosive potential. This is termed the "hungry water" phenomenon (Kondolf 1997). Incised channels often have reduced fish habitat and may reduce the river's ability to recharge groundwater. Levees, by reducing flooding, also reduce a river's role in depositing new alluvium on agricultural lands. This may have long-term adverse impacts on soil fertility. Levees have been removed along some reaches of the Sacramento and Cosumnes Rivers in recent years, restoring some of the older river dynamics.

Although dams are important tools in flood control and the impoundment of water for agriculture, they also can lead to changes in seasonal flow patterns. These changes can enable the invasion of exotic species by interfering with the life cycles of native plants and wildlife that are adapted to the river's natural cycles of flood and drought. For example, adverse effects have been observed on the foothill yellow-legged frog (*Rana boylii*) and on Fremont's cottonwood (*Populus fremontii*).

It is important to minimize changes to natural stream flows, which have a particularly important impact during the dry season when there is little available water to begin with. Water diversions for vineyards and other uses affect the hydrology of coastal watersheds, although the consequences for freshwater fish and other organisms are not well understood. Much of coastal California has a Mediterranean climate, with most of the rainfall occurring during winter. This pattern results in high stream flows in winter and little surface water during summer, when agriculture needs it most. Seasonal extremes in stream flow may be exacerbated by agricultural water use and by the presence of water-impervious surfaces or the consolidation of surface flows by engineered drainage networks, including those associated with vineyard development. The resulting hydrologic changes—including depletion of both surface water and groundwater—may have important geomorphic and ecological impacts (Merenlender 2008; Merenlender et al. 2008).

Recent data indicate that direct pumping from streams from late spring to mid-autumn can impact the flow levels of upland creeks. One alternative to pumping water on demand during the dry season is to use winter rainfall stored in farm ponds—small, geographically dispersed impoundments. However, the storage of winter rainfall in farm ponds requires that the landowner have appropriative water rights, which are increasingly difficult to obtain. Winter storage of water away from the stream channels is in many cases preferable to summer pumping from or near streams, since pumping leaves less water in the stream during the dry season when it is most important for the health of aquatic organisms.

Some vineyard managers are exploring ways to minimize this sort of impact on dry-season groundwater levels and the flow rates of streams. These techniques include the use of small, off-channel impoundments (ponds) mentioned above, and the collaborative, coordinated management of vineyard irrigation needs to minimize their effect on groundwater and surface waters (Merenlender 2008; Merenlender, Deitch, and Feirer 2008). Coordinated timing of irrigation is not always possible, however, especially in the case of frost protection, when the need for water is typically both urgent and widespread among growers.

Conservation of riparian vegetation—plants located along a course or body of water—should be a high priority. Riparian vegetation is home to the highest level of biodiversity in most ecosystems, including native birds in vineyard-dominated landscapes of Sonoma County (Luther et al. 2008). Diverse plant composition and structure, combined with both biotic and abiotic components (water, rocks, logs, etc.), provide niches that can support a broad array of species in a relatively small area (Garrison, Giusti, and Standiford 1996). Table 11.1 demonstrates the relative importance of riparian areas to vertebrates. Conservation of riparian vegetation can be a challenging goal for any grower who is also managing for Pierce's disease. The plants that are the best breeding hosts for the sharpshooters (Homoptera: Cicadellidae) that vector the disease and for the causal bacterium *Xylella fastidiosa* would have to be removed and then replaced with nonhost species to maintain the structure and function of riparian corridors.

In most of California, extensive seasonal wetlands have been drained or replaced with small, permanent lakes or ponds. This not only removes sensitive plant and animal species outright, it also facilitates invasion by nonnative species that fare better in the modified environments and have further negative impact on native species. In some cases, artificial wetlands and impoundments that enable or are the result of agricultural water use may appear on a superficial level to mimic natural features of California, but that appearance may hide important differences. For example, due to patterns of impoundment, re-channeling, and release, the seasonal abundance of water in many areas may differ profoundly from previous, natural patterns.

Many artificial wetlands and ponds contain water during seasons when it is lacking in nearby systems that follow more natural patterns, and this greater abundance can augment wildlife habitat. Moreover, widespread and rapid fluctuations in water levels caused by the drawdown and refill of reservoirs may inhibit the development of native emergent and riparian vegetation, the terrestrial plants naturally associated with creeks, streams, rivers, ponds, and lakes. Concrete-lined reservoirs may not be able to sustain emergent and riparian vegetation, even under the best of circumstances. Emergent and riparian plants are important in the ecology of insects, fish, amphibians, and other fauna.

Proper management of ponds, canals, natural wetlands, and streams requires that the manager take into account adjacent wildlands and their capacity to allow animals to survive during the nonbreeding season. Undisturbed areas are requisites of sensitive species such as native fish, amphibians, and reptiles. In California, likely target species for conservation and restoration in vineyards include but are not limited to reptiles such as western pond turtle (*Clemmys marmorata*) and several subspecies of common garter snake (*Thamnophis sirtalis*); amphibians such as California red-legged frog (*Rana aurora draytonii*), northern red-legged frog (*Rana aurora*), western toad (*Bufo boreas*), California tiger salamander (*Ambystoma californiense*), western newt (*Taricha torosa*), and rough-skinned newt (*Taricha granulosa*); and insects, including "charismatic species" (those known more for their visual beauty than for their pest behavior) such as damselflies and dragonflies (Odonata).

There is special interest in managing irrigation reservoirs to enhance the reproduction and survival of native amphibians. Permanent ponds often harbor introduced bullfrogs, nonnative fish, nonnative crayfish, and native dragonflies that interfere with the reproduction of several native amphibians. By draining farm ponds in November, a grower would reduce the occurrence of these predators so more native frogs and salamanders would reproduce. Native fish such as three-spined stickleback (*Gasterosteus aculeatus*) can be used for mosquito control and are thought to be less damaging to native amphibians than the commonly used, but introduced, mosquitofish (*Gambusia affinis*). Establishment of aquatic and emergent vegetation (e.g., bulrushes [tules] [*Scirpus* spp.] or cattails [*Typha* spp.]) can provide food and cover for reproducing and metamorphosing amphibians, reducing both predation and desiccation. One study of farm ponds in southeastern Australia found the highest degree of frog diversity to be in ponds with abundant emergent vegetation, a large percentage of ground cover in the riparian zone, and a large percentage of cover by native vegetation within 1 kilometer of the pond.

In vineyards in California, the following considerations apply. They may also be adapted for other regions.

- Reservoir design should include a section that is at least 1.2 meters deep, with adjoining emergent vegetation such as bulrushes or cattails to provide refuge from heat and from predators (Figures 11.2 and 11.3).

- Design should include a shallow end that is bare or has low-growing plants that will permit basking.

- Add native emergent and riparian vegetation to edges.

- Include adjoining groves of riparian plants with moist, marshy sections (via drip line).

- Remove and subsequently exclude nonnative Centrarchidae (bass and sunfish).

- Drain pond every 2 years to reduce use by non-native North American bullfrog (*Rana catesbaiana* [=*Lithobates catesbeianus*]).

- If pond is to be drained only rarely, use Sacramento perch (*Archoplites interruptus*), three-spined stickleback (*Gasterosteus aculeatus*), or other native fish, or rely on native insects (Hemiptera and Odonata) for mosquito control.

Figure 11.2. Pond adjoining vineyard in Oregon. Note aquatic and emergent vegetation, which are important in enhancing amphibian diversity and abundance. *Photo* by Rebecca M. Sweet.

Figure 11.3. Pond adjoining vineyard in Oregon. Here, the aquatic vegetation appears limited to algal mats, and the emergent and riparian vegetation are limited by close mowing. These conditions may in turn limit amphibian abundance and diversity. *Photo* by Rebecca M. Sweet.

- High mowing and reduced tillage of green cover will make the vineyard understory more hospitable to amphibians and reptiles.

Many amphibians and reptiles use rodent burrows for shelter, enabling aestivation or hibernation. Rodents, such as voles (e.g. *Microtus californicus*), pocket gophers, and ground squirrels (e.g. *Spermophilus beecheyi*), can also be pests, however. This may make for a difficult balancing act. The use of PVC pipes as artificial burrows has been proven as a successful technique for enhancing Japanese tree frog (*Hyla japonica*) in Japanese rice fields. This same type of artificial burrow should be assessed for its effectiveness in conserving other amphibians and reptiles.

Even ornamental ponds and fountains such as those near tasting rooms of wineries can be designed, retrofitted, or managed to accommodate native amphibians. Many species of toads, frogs, and salamanders may require something like a ramp to allow them to enter and exit such water features. The ramps can take the form of stones, pieces of wood, or living plants. One exception is Pacific tree frog (Pacific chorus frog, *Hyla [Pseudacris] regilla*), which has suction cups on its toes, which allow it to scale most vertical surfaces. The presence of aquatic, emergent, and riparian vegetation, as well as the annual drainage of ponds in late fall, may be important for enhancing amphibian reproduction and survival.

Cover crops. Cover crops are widely used in vineyards for erosion control and nitrogen fixation.

The makeup of a cover crop mixture is important, because the presence of multiple species can provide functional redundancies and complementarities. For example, functional redundancy occurs when multiple legumes are used in a mix: if one species grows poorly, another may compensate, providing backup. Functional complementarities can be obtained by seeding grasses and legumes together: grasses are often more efficient at scavenging soil nitrate, whereas legumes fix atmospheric nitrogen. These cases highlight key aspects of functional biodiversity.

Cover crops may be grazed by or their seeds may be fed upon by herbivores, including insects and other arthropods. These herbivores in turn serve as prey for carnivorous species. Several winter-annual cover crops appear to harbor a higher density of aphids, plant bugs, and associated predators than does resident vegetation, although no formal comparisons have been published. Bean aphid (*Aphis fabae*), cowpea aphid (*Aphis craccivora*), and pea aphid (*Acyrthosiphon pisum*) occur on winter-annual legumes, oat-bird cherry aphid (*Rhopalosiphum padi*) infests cereals, and cabbage aphid (*Brevicoryne brassicae*) and mustard aphid (*Lipaphis erysimi*) occur on mustards. Tarnished plant bug (*Lygus* spp.) and Norwegian plant bug (*Calocoris norvegicus*) occur on vetches, burr medic, and clovers (with the exception of subterranean clovers). Black grass bugs (*Labops* spp.) occur on some grasses, especially on California's North Coast. Most arthropods on these cover crops are abundant from late March to early May, coinciding with the grapevine bloom period, and they may be important foods for birds and predatory insects.

For at least part of the year, many winegrape growers use resident vegetation as a cover crop. In North Coast counties of California (Napa, Lake, Sonoma, and Mendocino Counties), winter-annual resident vegetation is used to protect the vineyard soil from erosion during the most intense winter rains. Tillage by mid-March and at intervals thereafter is intended to provide vegetation-free middles until fall rains enable the winter-annual resident vegetation seeds to germinate. Given this sort of schedule, only early maturing, low-growing, low-biomass herbaceous plants are able to complete their life cycles. Species that are commonly encountered in such vineyards include common chickweed (*Stellaria media*), henbit (*Lamium amplexicaule*), scarlet pimpernel (*Anagallus arvensis*), shepherd's-purse (*Capsella bursa-pastoris*), annual bluegrass (*Poa annua*), rattail fescue (*Vulpia myuros*), wild oat (*Avena fatua*), and darnel (*Lolium tementulum*). None of these species is native to California. Where tillage can be postponed till late March or early April, self-reseeding stands of field mustard (*Brassica campestris*) may be managed. This species confers the classic "golden vineyard" look by early spring (Figure 11.4).

Most vineyard cover cropping entails seeding winter-annual grasses, legumes, and mustards, either singly or in mixes. Large-seeded "plow-down" mixes typically consist of bell bean (*Vicia faba*) (Figure 11.5), common vetch (*Vicia sativa*) (Figure 11.6), and field pea (*Pisum sativum* ssp. *arvense*) (Figure 11.7),

---

Sidebar 11-D

REGIONAL BIODIVERSITY

California is a biologically diverse and ecologically rich state. The combination of mountains, oceans, rivers, streams, valleys, deserts, marshes, forests, woodlands, and grasslands provides a combination of habitat types and ecotones (zones of transition between two different ecosystems, e.g., where the sea meets the land) found in few places in the world. This assemblage of habitat types gives California an abundance of unique and diverse soils, plants, and animals.

Each region of California has unique ecological communities and histories that have led to the current condition of those habitats and communities. Any attempt to address the maintenance and enhancement of biological diversity on a given property must take these factors into account:

1. the geographical location of the property

2. the condition of the surrounding area and how it may have been altered from historic times

3. the design and management options available to retain, recruit, and restore important ecological elements

*   **Interior valleys** — These include the Sacramento and the San Joaquin Valleys and contain deep and fertile alluvial soils resulting from great river systems. Though the primordial oak forests have been dramatically altered since European settlement, the richness and depth of the soils make regeneration and remediation very achievable. The most glaring adverse legacy condition of the interior valley regions is the absence of native trees, e.g., valley oak *(Quercus lobata)*, California sycamore *(Platanus racemosa),* and willows *(Salix* spp.). A simple and relatively easily accomplished project in these soil-rich environments to improve biological diversity would be to retain and recruit native trees, either as individuals or in hedgerows, to improve the structural diversity of many areas.

*   **Foothills** — Development continues to fragment habitats in many foothill zones. Retaining contiguous tracts that harbor wild plants will assist in the long-term maintenance of functional systems. Many foothill zones still have remnant patches (sometimes very large) of oak woodlands and mixed oak-conifer forests. The best approach to retaining functional habitat and thus retaining the biological diversity of the region is to avoid or minimize impacts to the structure and composition of these existing vegetation communities.

*   **Coastal valley** — Most coastal valleys have been highly altered. Most are relatively small and most have highly fertile soils. Like the Central Valley, coastal valleys have been altered dramatically since pre-European times. Stream corridors in coastal valley settings are subject to regulatory constraints intended to aid salmonid fish populations. It is important to retain and promote streamside vegetation wherever possible, as this has in most cases been degraded by past and ongoing management decisions. Riparian and emergent vegetation provide habitat for terrestrial and aquatic animal species, maintain a cool, moist understory atmosphere, provide shade to regulate water temperatures, and ensure leaf fall that stimulates nutrient cycling within the system. Retention and recruitment of tree structure, whether tree-by-tree or in groups, can improve the structural diversity of many areas.

*   **Coastal mountains** — Hardwood forests are generally distributed in patchy patterns in the coastal mountains. The best approach to retaining biological diversity is to minimize or avoid any disturbance of these patches. As with coastal valleys, the retention of streamside vegetation is important for both terrestrial and aquatic animal species as well as for keeping soil subsidence into the stream (sedimentation) to a minimum.

Figure 11.4. Field mustard *(Brassica campestris)* can be used as a self-reseeding winter-annual cover crop. *Photo* by Robert L. Bugg.

Figure 11.5. Bell bean *(Vicia faba)*; dark spot on stipule is extrafloral nectary. *Photo* by Robert L. Bugg.

Figure 11.6. Common vetch *(Vicia sativa)*; dark spot on stipule is extrafloral nectary. *Photo* by Robert L. Bugg.

Figure 11.7. A plow-down mix of large-seeded winter-annual cover crops, including barley *(Hordeum vulgare)*, common vetch *(Vicia sativa)*, and field pea *(Pisum sativum* ssp. *arvense)*. *Photo* by Glenn T. McGourty.

in combination with oat *(Avena sativa)*. Small-seeded winter-annual mixes are usually managed without tillage, and often include bur medic *(Medicago polymorpha)*, various annual clovers *(Trifolium* spp.) (Figures 11.8, 11.9, and 11.10), and an annual grass. The annual medic-clover mixes are expected to volunteer (regenerate on their own) for several successive years.

On soils of medium to high fertility, many organic grape growers use some perennial grasses managed without tillage, at least in alternate middles, or alleys. These grasses support foot and vehicular traffic better than legumes do and produce less dust from traffic than tilled soil. Annual legumes, including common vetch, and perennial legumes such as Ladino clover or white Dutch clover *(Trifolium pratense)* (Figure 11.11) can be used in combination with perennial grasses

and are believed to offset competition with the vines for soil nitrogen. Birdsfoot trefoil *(Lotus corniculatus)* (Figure 11.12) is slow to establish but spreads reasonably well because it has hard seed. Strawberry clover *(Trifolium fragiferum)* is usually avoided because this perennial clover attracts and sustains high densities of pocket gophers (e.g. *Thomomys bottae)*. These are especially problematic in young vineyards.

Perennial grasses can be categorized as low impact, medium impact, or high impact, depending on their tendency to compete with the vines for water or nutrients. In California, this is called "de-vigoration." Low impact grasses include the fine fescues, sheep fescue, and red fescues. California native fine fescues (e.g. *Festuca rubra, Festuca idahoensis)* have lower seedling vigor than several of the introduced, improved varieties. Soil crusting, freeze-thaw and wetting-drying

Figure 11.8. Persian clover *(Trifolium resupinatum)*. *Photo* by Robert L. Bugg.

Figure 11.9. Subterranean clover *(Trifolium subterraneum)*. *Photo* by Robert L. Bugg.

Figure 11.10. Subterranean clover *(Trifolium subterraneum)* interspersed with bur medic *(Medicago polymorpha)* and tansy phacelia *(Phacelia tanacetifolia)*. *Photo* by Glenn T. McGourty.

Figure 11.11. A mixture of perennial and annual cover crops, including 'Hounddog' tall fescue *(Festuca arundinacea)*, crimson clover *(Trifolium incarnatum)*, California poppy *(Eschscholtzia californica)*, and tansy-leafed phacelia *(Phacelia tanacetifolia)*. *Photo* by Robert L. Bugg.

Figure 11.12. Birdsfoot trefoil *(Lotus corniculatus)*. *Photo* by Robert L. Bugg.

cycles all can kill seedlings, and low seedling vigor may exacerbate these losses. Medium-impact grasses include turf selections of perennial ryegrass such as Manhattan II, which is a bunchgrass of Mediterranean origin. High-impact grasses include tall fescue (*Festuca arundinacea*) varieties such as Hounddog, hard fescue (*Festuca brevipila*), Berber orchard grass (*Dactylis glomerata*), and tall-statured California native grasses such as blue wild rye (*Elymus glaucus*) and purple needlegrass (*Nassella pulchra*).

Summer-annual cover crops are rarely used in California vineyards, with the most common exception being buckwheat (*Fagopyrum esculentum*), a spring-sown summer annual. Buckwheat is a floral nectar source for numerous parasitic and predatory insects, including some that attack grape pests (English-Loeb et al. 2003).

California native grasses and wildflowers rarely volunteer in vineyards so seldom occur in a spontaneous green cover. One exception is redmaids (*Calandrinia ciliata*), a widespread winter-annual forb with spectacular magenta blossoms. In some cases, native plants are seeded, sometimes in combination with nonnative plants. In a 2-year study in Oregon, Sweet (2006) found that seeded native plants were more expensive than other options but performed well in terms of enabling vine growth and grape yield and quality.

Here are three approaches (with partial overlap) to cover cropping that predominate in Californian vineyards:

1. *Annual Cover Cropping with Tillage (Plow-Down Approach).* This approach typically uses mixtures of large-seeded legumes such as bell bean (*Vicia faba*) (Figure 11.5), common vetch (*Vicia sativa*) (Figure 11.6), and field pea (*Pisum sativum* ssp. *arvense*) (Figure 11.7) in combination one or more cereal grasses, typically a robust oat (*Avena sativa*) variety (e.g., cv. Ogle or Swan). In cases where nitrogen inputs are not needed, early maturing mustard (e.g., field mustard [*Brassica campestris*] [Figure 11.4]) or a cereal grass (e.g., oat, cereal rye, or barley) may be used. A key aim for some variants of this approach is moisture conservation. This is especially true for dry-farmed vineyards.

A typical management scheme involves seeding with a drill during October or November, followed immediately by compost application. The cover crop is flail- or rotary-mowed in April or May, then disked under while the soil is still moist. Immediately thereafter, shanks with duckfoot sweeps, spring cultivators, and weed knives are used to take out summer weeds such as lambsquarters, prickly lettuce (*Lactuca serriola*), and pigweed (*Amaranthus retroflexus*). The surface is finished with a ringroller, creating a dust mulch that breaks capillarity, thus decreasing evaporative loss of soil moisture.

2. *Alternate-Middle Management with Plow-Down Annual Cover Crops and Mown Self-Reseeding Annual Cover Crops* (Figures 11.13, 11.14, and 11.15). This approach uses plow-down legume-cereal mixes like those described above, alternated spatially with small-seed, low-growing, self-reseeding legumes (e.g., annual clovers and medics, sometimes combined with annual grasses and with common vetch). This approach is used on intermediate- and high-fertility sites. Data on transverse conduction of water by grapevines (Smart et al. 2005) suggest that the grapevines may be able to balance for the effects of differing water availabilities in alternate middle alleys.

A common management sequence is to seed the plow-down annual cover crop and add compost as above. A variant of this general approach would involve reduced tillage and the use of a no-till seeder (slit seeder) to sow the maximum biomass mix. The cover crop would then be mowed down in mid- to late spring. The self-reseeding mixture is seeded in alternate middles. The plow-down middles may be treated as above, whereas the self-reseeding mix is managed with periodic mowing. After 3 years, growers switch so that the cover cropping regimes are alternated temporally as well as spatially. This rotation is thought to reduce soil compaction and pathogen buildup, but there are no experimental data documenting these effects.

3. *No-Till or Reduced-Tillage Management.* This approach provides several options for cover cropping in the middles and may use (a) self-reseeding annual legumes, (b) perennial grasses, (c) perennial legumes, or (d) a mixture of two or more of these categories. For weed control in vine rows, this approach may rely on under-vine tillage with articulating in-row cultivators (e.g., the Clemens implement) or under-vine flaming. Legumes may fade out over time,

Figure 11.13. The California native annual forb baby blue eyes *(Nemophila menziesii),* grown as a vineyard cover crop. Field mustard *(Brassica campestris)* occurs in the adjoining alley (=middle). *Photo by Glenn T. McGourty.*

Figures 11.14 and 11.15. Alternate middle management with fall-sown crimson clover *(Trifolium incarnatum)* and spring-sown buckwheat *(Fagopyrum esculentum). Photos by Glenn T. McGourty.*

though no formal data are available on this. This observed phenomenon raises the possibility of shifting the mix subtly through time to maintain enough legumes to meet the nitrogen needs of the system. This can be achieved by no-till drilling new legume seeds into the existing swards. To accommodate ground-nesting birds, it is important that vineyard managers carefully time spring mowing and tillage. Otherwise they may bring on an "attractive sink" scenario, where birds such as western meadowlark or red-winged blackbird are attracted to nest in a standing cover crop that is later mowed, destroying the nests and nestlings. Mowing in late February or March to keep cover crops low and unattractive to birds during the nest selection period may be part of the solution.

Botta's pocket gopher *(Thomomys bottae)* is very damaging to young vines but less so to mature ones. In mature vineyards and along vineyard borders, consider using traps to remove pocket gophers while leaving their burrows intact. Rodent burrows provide aestivation and hibernation sites for amphibians and reptiles. Cover-cropping regimes can be designed to help manage these animals while minimizing their damage to vines: perennial legumes, especially strawberry clover *(Trifolium fragiferum),* appear to harbor the highest densities of this rodent, and annual grasses the lowest. Alternatively, growers can provide amphibian and reptile habitat in the form of rocks, logs, and woodpiles on the edges of the vineyard (see Whisson and Giusti 1998).

Hedgerows. Hedgerows, whether large or small, serve as screens and barriers. A hedgerow made up of shrubs will be low enough to let sunlight reach the vineyard while helping reduce dust drift from adjacent roads and screening vineyard operations from the view of neighbors. Small islands of trees are aesthetically pleasing and can provide places for birds to roost without creating large shade patterns or competition for vines. Some growers limit most of their edge plantings to herbaceous plants rather than hedgerows, since they can be mowed seasonally to reduce fire danger. Herbaceous plants are still helpful, especially if they help reduce erosion along watercourses or filter silt and nutrients from water leaving a vineyard.

Some growers incorporate low-maintenance fruit and nut trees such as persimmons, pomegranates, olives, jujubes, medlars, almonds, and gooseberries. These provide fruit for people and wildlife. Some ornamental plants are exceptionally attractive to bees and wasps, such as members of the mint family and the carrot family. These include culinary herbs such as dill, fennel, lemon verbena, mint, oregano, rosemary, sage, and thyme.

Often, growers will use native plants in hedgerows, partly compensating for losses of native flora to agricultural and urban development. The plants may need irrigation at first, but once established they can often survive well on rainfall alone. One key is to choose plants that are specifically adapted to your region. Hedgerows alongside many vineyards consist simply of the edges where the grower has decided to tolerate the growth of volunteer plants, including trees, shrubs, forbs, and grasses.

For the vineyardist designing a hedgerow that will harbor beneficial insects, it is useful to plant multiple species that, together, provide blooms through most of the year. Choose species that do not harbor diseases or pests of winegrapes. For example, avoid *Ceanothus* since it may harbor Eutypa and avoid blackberries in areas where Pierce's disease is common. Irrigate during establishment (a drip system attached to the vineyard's existing drip system will work well), control weeds, and provide protection from rabbits and deer when trees and shrubs are small. Periodic pruning and mowing often help to rejuvenate plants and encourage bloom and vegetative growth. (For details on management, see Earnshaw 2004.)

Fencing. Fences can act as an effective barrier to both target and nontarget animals. Fences are effective at keeping deer from feeding on grapevines, but they also affect the movement of nonpest species, forcing them to take paths that may lead them to their death by means of predation or collision with motor vehicles. Though fences may be needed for some vineyards, you can reduce their impact on nontarget wildlife by leaving some space at the bottom of the fence to permit the passage of animals that are smaller than deer. You can also limit the fenced area to the vineyard alone, rather than the entire property. The use of fencing "blocks," with unfenced corridors between the blocks, is another approach. In both cases, the aim is to increase permeability to wildlife and maintain wildlife access to water, food, cover, and neighboring tracts.

The use of wildlife-friendly fencing wherever possible minimizes the degree of habitat fragmentation. In North America, the currently accepted wildlife-friendly fence standards specify three strands of smooth wire, with the bottom strand at least 16 inches above the ground, the second strand at 24 inches, the third wire at 32 inches, and a horizontal pole on top at 40 inches. The top pole provides an important visual cue so wildlife can see the fence and avoid getting tangled in the top wire. Generally, the more visible the materials are, the better, so wood, recycled plastic, and PVC are better options than wire. Do not fence across or along creeks, as these waterways are particularly attractive to wildlife. Do not fence-in deer and other wildlife; wildlife can overgraze the available vegetation in an enclosed area, and they should not be isolated from surrounding populations. If you need to exclude deer from a vineyard, a fence with staggered pickets 5 feet high is preferable to a 6- to 8-foot-high woven wire fence that reaches all the way to the ground.

SUMMARY

In the face of widespread development and population growth, Californians face an ongoing challenge in working to conserve and restore natural resources, including native biodiversity. Even seemingly subtle changes in management can have a large impact. For vineyard and winery owners and operators, many opportunities for conservation and restoration arise as part of making decisions on landscaping, cover cropping, fencing, irrigation reservoirs, and riparian zones. There may also be opportunities to enhance conservation efforts by coordinating and collaborating with managers of adjoining lands, both private and public, to achieve landscape-scale benefits for many species of plants and animals. Such collaborations could span urban, suburban, agricultural, and wild lands. Collective approaches, though more difficult to plan and implement than individual efforts, may in the long run be more fruitful.

An individual winegrower can maintain and enhance the biological diversity in a vineyard and can also benefit from the input of private ecological consultants. However, a piecemeal approach to biodiversity enhancement will not give meaningful results. To stop the loss of biodiversity in California's agricultural lands, landowners and public land managers and regulators will have to make a cooperative effort. Only landscape-scale measures to maintain and restore California's natural and agricultural ecosystems will ensure that future generations are able to enjoy nature and continue to rely on its goods and services.

REFERENCES/RESOURCES

Adams, M. J. 2000. Pond permanence and the effects of exotic vertebrates on anurans. Ecological Applications 10:559–568.

Adler, S. 2003. California vineyards and wildlife habitat. Sacramento: California Association of Winegrape Growers.

Baker, J. M. R., and T. R. Halliday. 1999. Amphibian colonization of new ponds in an agricultural landscape. Herpetological Journal 9(2):55–63.

Berge, A., M. Delwiche, W. P. Gorenzel, and T. Salmon. 2007. Bird control in vineyards using alarm and distress calls. American Journal of Enology and Viticulture 58(1):135–143.

Blaustein, A. R., and J. M. Kiesecker. 2001. Complexity in conservation: Lessons from the global decline of amphibian populations. Ecology Letters 5:597–608.

Bogner, F. X. 1999. Empirical evaluation of an educational conservation programme introduced in Swiss secondary schools. International Journal of Science Education 21:1169–1185.

Bossard, C. C., J. M. Randall, and M. C. Hoshovsky. 2000. Invasive plants of California's wildlands. Berkeley: University of California Press.

Boudreau, G. W. 1972. Factors related to bird depredations in vineyards. American Journal of Enology and Viticulture 23(2):50–53.

Boutin, C., K. E. Freemark, and D. A. Kirk. 1999. Spatial and temporal patterns of bird use of farmland in southern Ontario. Canadian Field-Naturalist 113:430–460.

Brotons, L., M. Monkkonen, and J. L. Martin. 2003. Are fragments islands? Landscape context and density-area relationships in boreal forest birds. American Naturalist 162:343–357.

Brown, C. S., and R. L. Bugg. 2001. Effects of established perennial grasses on introduction of native forbs in California. Restoration Ecology 9:38–48.

Bugg, R. L., J. H. Anderson, C. D. Thomsen, and J. Chandler. 1998. Farmscaping: Restoring native biodiversity to agricultural settings. In C. H. Pickett and R. L. Bugg, eds., Enhancing biological control: Habitat management to promote natural enemies of agricultural pests. Berkeley: University of California Press. Pp. 339–374

Bugg, R. L., C. S. Brown, and J. H. Anderson. 1997. Restoring native perennial grasses to rural roadsides in the Sacramento Valley of California: Establishment and evaluation. Restoration Ecology 5:214–228.

Bugg, R. L., G. McGourty, M. Sarrantonio, W. T. Lanini, and R. Bartolucci. 1996. Comparison of thirty-two cover crops in an organic vineyard on the north coast of California. Biological Agriculture and Horticulture 13:65–83.

Bugg, R. L., S. C. Phatak, and J. D. Dutcher. 1990. Insects associated with cool-season cover crops in southern Georgia: Implications for pest control in the truck-farm and pecan agroecosystems. Biological Agriculture and Horticulture 7:17–45.

Bugg, R. L., and P. C. Trenham. 2003a. Agriculture affects amphibians (Part 1): Climate change, landscape-scale dynamics, hydrology, mineral enrichment of water. Sustainable Agriculture (newsletter of UC SAREP), Spring 2003, 15(1):12–15.

———. 2003b. Agriculture affects amphibians (Part 2): Pesticides, fungi, algae, higher plants, fauna, management recommendations. Sustainable Agriculture (newsletter of UC SAREP), Fall 2003, 15(2):8–11.

Bugg, R. L., and M. Van Horn. 1998. Ecological soil management and soil fauna: Best practices in California vineyards. In R. Hamilton, L. Tassie, and P. Hayes (eds.), Proceedings of the Viticulture Seminar: Viticultural Best Practice, Mildura Arts Centre, 1 August, 1997, Mildura, Victoria, Australia. Australian Society for Viticulture and Oenology, Inc., Adelaide, South Australia, Australia. Pp. 23–34.

Bugg, R. L., F. L. Wäckers, K. E. Brunson, S. C. Phatak, and J. D. Dutcher. 1990. Tarnished plant bug (Hemiptera: Miridae) on selected cool-season leguminous cover crops. Journal of Entomological Science 25(3):463–474.

Celette, F., J. Wery, E. Chantelot, J. Celette, and C. Gary. 2005. Belowground interactions in a vine (*Vitis vinifera* L.)–tall fescue (*Festuca arundinacea*

Shreb.) intercropping system: Water relations and growth. Plant and Soil 276:205–217.

Céréghino, R., A. Ruggiero, P. Marty, and S. Angélibert. 2008. Biodiversity and distribution patterns of freshwater invertebrates in farm ponds of a southwestern French agricultural landscape. Hydrobiologia 597:43–51.

Chamberlain, D. E., J. D. Wilson, and R. J. Fuller. 1999. A comparison of bird populations on organic and conventional farm systems in southern Britain. Biological Conservation 88:307–320.

Chan-Mcleod, A. C. A., and A. Moy. 2007. Evaluating residual tree patches as stepping stones and short-term refugia for red-legged frogs. Journal of Wildlife Management 71(6):1836–1844.

Coipel, J., B. R. Lovelle, C. Sipp, and C. Van Leeuwen. 2006. "Terroir" effect, as a result of environmental stress, depends more on soil depth than on soil type (Vitis vinifera L. cv. Grenache noir, Cotes du Rhone, France, 2000). Journal International des Sciences de la Vigne et du Vin 40(4):177–185.

Conover, M. R. 1998. Perceptions of American agricultural producers about wildlife on their farms and ranches. Wildlife Society Bulletin 26:597–604.

Costello, M. J., and K. M. Daane. 1998. Influence of ground cover on spider populations in a table grape vineyard. Ecological Entomology 23(1):33–40.

———. 2003. Spider and leafhopper (Erythroneura spp.) response to vineyard ground cover. Environmental Entomology 32(5):1085–1098.

Curtis, J. M. R., and E. B. Taylor. 2004. The genetic structure of coastal giant salamanders (Dicamptodon tenebrosus) in a managed forest. Biological Conservation 115:45–54.

Dallman, P. R. 1998. Plant life in the world's Mediterranean climates: California, Chile, South Africa, Australia, and the Mediterranean Basin. Berkeley: University of California Press.

Davidson, C., H. B. Shaffer, and M. R. Jennings. 2001. Declines of the California red-legged frog: Climate, UV-B, habitat, and pesticides hypotheses. Ecological Applications 11(2):464–479.

Davies, B. R., J. Biggs, P. J. Williams, J. T. Lee, and S. Thompson. 2008b. A comparison of the catchment sizes of rivers, streams, ponds, ditches and lakes: Implications for protecting aquatic biodiversity in an agricultural landscape. Hydrobiologia 597:7–17.

Davies, B., J. Biggs, P. Williams, M. Whitfield, P. Nicolet, D. Sear, S. Bray, S. Maund. 2008a. Comparative biodiversity of aquatic habitats in the European agricultural landscape. Agriculture, Ecosystems, and Environment 125:1–8.

Doubledee, R. A., E. B. Muller, and R. M. Nisbet. 2003. Bullfrogs, disturbance regimes, and the persistence of California red-legged frogs. Journal of Wildlife Management 67(2):424–438.

Earnshaw, S. 2004. Hedgerows for California agriculture: A resource guide. Davis, California: Community Alliance with Family Farmers. www.caf.org/programs/farmscaping/Hedgerow.pdf.

English-Loeb, G, M. Rhainds, T. Martinson, T. Ugine. 2003. Influence of flowering cover crops on Anagrus parasitoids (Hymenoptera : Mymaridae) and Erythroneura leafhoppers (Homoptera: Cicadellidae) in New York vineyards. Agricultural and Forest Entomology 5(2):173–181.

Fahrig, L., J. H. Pedlar, S. E. Pope, P. D. Taylor, and J. F. Wegner. 1995. Effect of road traffic on amphibian density. Biological Conservation 73:177–182.

Fiehler, C. M., W. D. Tietje, and W. R. Fields. 2006. Nesting success of Western Bluebirds (Sialia mexicana) using nest boxes in vineyard and oak-savannah habitats of California. Wilson Journal of Ornithology 118(4):552–557.

Fluetsch, K. M., and D. W. Sparling. 1994. Avian nesting success and diversity in conventionally and organically managed apple orchards. Environmental Toxicology and Chemistry 13:1651–1659.

Franklin, J. F. 1988. Structural and functional diversity in temperate forests. In E. O. Wilson, ed., Biodiversity. Washington, D.C.: National Academy Press. Pp. 166–175.

Freemark, K. E., and D. A. Kirk. 2001. Birds on organic and conventional farms in Ontario: Partitioning effects of habitat and practices on species composition and abundance. Biological Conservation 101:337–350.

Gall, G. A. E., and G. H. Orians. 1992. Agriculture and biological conservation. Agriculture Ecosystems and Environment 42(1–2):1–8.

Gall, G. A. E., and M. Staton. 1992. Integrating conservation biology and agricultural production: Conclusions. Agriculture Ecosystems and Environment 42:217–230.

Garrison, B., G. A. Giusti, and R. B. Standiford. 1996. Oaks and Habitats of the Hardwood Range. In R. B. Standiford (ed.), Guidelines for managing California's hardwood rangelands. Oakland: University of California, Division of Agriculture and Natural Resources, Publication 3368. Pp. 8–17.

Gibbs, J. P. 1998. Distribution of woodland amphibians along a forest fragmentation gradient. Landscape Ecology 13:263–268.

Giusti, G. A., B. Garrison, and E. L. Fitzhugh. 1996. Developing recreational sources of income from oak woodlands. In R. B. Standiford (ed.), Guidelines for managing California's hardwood rangelands. Oakland: University of California, Division of Agriculture and Natural Resources, Publication 3368. Pp. 68–77.

Giusti, G. A., R. J. Keiffer, and C. E. Vaughn. 2003. The bird community of an oak woodland stream. California Fish and Game 89(2):72–80.

Giusti, G. A., T. Scott, B. Garrison, and K. Shaffer. 1996. Oak woodland wildlife ecology, native plans and habitat relationships. In R. B. Standiford (ed.), Guidelines for managing California's hardwood rangelands. Oakland: University of California, Division of Agriculture and Natural Resources, Publication 3368. Pp. 34–50.

Golet, G. H., M. D. Roberts, R. A. Luster, G. Werner, E. W. Larsen, R. Unger, and G. G. White. 2006. Assessing societal impacts when planning restoration of large alluvial rivers: A case study of the Sacramento River Project, California. Environmental Management 37(6):862–879.

Gong, P., S. A. Mahler, G. S. Biging, and D. A. Newburn. 2003. Vineyard identification in an oak woodland landscape with airborne digital camera imagery. International Journal of Remote Sensing 24(6):1303–1315.

Goodsell, J. A., and L. B. Kats. 1999. Effect of introduced mosquito fish on pacific tree frogs and the role of alternative prey. Conservation Biology 13:921–924.

Haddad, B. M. 2003. Property rights, ecosystem management, and John Locke's labor theory of ownership. Ecological Economics 46(1):19–31.

Hazell, D., R. Cunningham, D. Lindenmayer, B. Mackey, and W. Osborne. 2001. Use of farm dams as frog habitat in an Australian agricultural landscape: Factors affecting species richness and distribution. Biological Conservation 102:155–169.

Hazell, D., J. M. Hero, D. Lindenmayer, and R. Cunningham. 2004. A comparison of constructed and natural habitat for frog conservation in an Australian agricultural landscape. Biological Conservation 119(1):61–71.

Hazell, D., W. Osborne, and D. Lindenmayer. 2003. Impact of post-European stream change on frog habitat: Southeastern Australia. Biodiversity and Conservation 12(2):301–320.

Heaton, E., R. Long, C. Ingels, and T. Hoffman. 2008. Songbird, bat and owl boxes: Vineyard management with an eye toward wildlife. Oakland: University of California, Division of Agriculture and Natural Resources, Publication 21636.

Hilty, J. A., 2001. Use of riparian corridors by wildlife in the oak woodland vineyard landscape. Ph.D. Dissertation. Department of Environmental Science, Policy, and Management. University of California, Berkeley. 268 pp.

Hilty, J. A., W. Z. Lidicker Jr., A. Merenlender, and A. P. Dobson. 2006. The science and practice of linking landscapes for biodiversity conservation, 1st ed. Washington, D.C.: Island Press. 344 pp.

Hilty, J. A., and A. Merenlender. 2000. Faunal indicator taxa selection for monitoring ecosystem health. Biol. Conserv. 92:185–197.

———. 2004. Use of riparian corridors and vineyards by mammalian predators in northern California. Conservation Biology 18(1):126–135.

Hobbs, R. J., and C. J. Yates. 2003. Impacts of ecosystem fragmentation on plant populations: Generalising the idiosyncratic. Australian Journal of Botany 51:471–488.

Imbeau, L., P. Drapeau, and M. Mokkonen. 2003. Are forest birds categorised as "edge species" strictly associated with edges? Ecography 26:514–520.

Imhoff, D. 2003. Farming with the wild: Enhancing biodiversity on farms and ranches. San Francisco: Sierra Club Books.

Ingels, C. A., R. L. Bugg, G. T. McGourty, and L. P. Christensen, eds. 1998. Cover cropping in vineyards: A grower's handbook. Oakland: University of California, Division of Agriculture and Natural Resources, Publication 3338.

Joly, P., C. Miaud, A. Lehmann, and O. Grolet. 2001. Habitat matrix effects on pond occupancy in newts. Conservation Biology 15:239–248.

Joly, P., C. Morand, and A. Cohas. 2003. Habitat fragmentation and amphibian conservation: Building a tool for assessing landscape matrix connectivity. Comptes Rendus Biologies 326: S132–S139 Suppl. 1.

Kiesecker, J. M., and A. R. Blaustein. 1998. Effects of introduced bullfrogs and smallmouth bass on microhabitat use, growth, and survival of native red-legged frogs (Rana aurora). Conservation Biology 12(4):776–787.

Kiesecker, J. M., A. R. Blaustein, and C. L. Miller. 2001. Potential mechanisms underlying the displacement of native red-legged frogs by introduced bullfrogs. Ecology 82(7):1964–1970.

Klik, A., J. Rosner, and W. Loiskandl. 1998. Effects of temporary and permanent soil cover on grape yield and soil chemical and physical properties. Journal of Soil and Water Conservation 53(3):249–253.

Knutson, M. G., W. B. Richardson, D. M. Reineke, B. R. Gray, J. R. Parmelee, and S. E. Weick. 2004. Agricultural ponds support amphibian populations. Ecological Applications 14(3): 669–684.

Kolozsvary, M. B., and R. K. Swihart. 1999. Habitat fragmentation and the distribution of amphibians: Patch and landscape correlates in farmland. Canadian Journal of Zoology 77:1288–1299.

Kondolf, G. M. 1997. Profile: Hungry water: Effects of dams and gravel mining on river channels. Environmental Management 21(4):533–551.

Kruess, A., and T. Tscharntke. 1994. Habitat fragmentation, species loss, and biological control. Science 264:1581–1584.

Lacher, T. E., Jr., R. D. Slack, L. M. Coburn, and M. I. Goldstein. 1999. The role of agroecosystems in wildlife biodiversity. In W. W. Collins and C. O. Qualset (eds.), Biodiversity in agroecosystems. Boca Raton: CRC Press. Pp. 147–165.

Lattin, J. D. 2006. Observations of selected true bugs (Hemiptera: Heteroptera) of the Pacific Northwest shrub-steppe zone. Western North American Naturalist 66(2):256–259.

Lawler, S. P., D. Dritz, T. Strange, and M. Holyoak. 1999. Effects of introduced mosquitofish and bullfrogs on the threatened California red-legged frog. Conservation Biology 13:613–622.

LeCoeur, D., J. Baudry, F. Burel, and C. Thenail. 2002. Why and how we should study field boundary biodiversity in an agrarian landscape context. Agriculture, Ecosystems, and Environment 89:23–40.

Lokemoen, J. T., and J. A. Beiser. 1997. Bird use and nesting in conventional, minimum-tillage, and organic cropland. Journal of Wildlife Management 61:644–655.

Lovett-Doust, J., M. Biernacki, R. Page, M. Chan, R. Natgunarajah, and G. Timis. 2003. Effects of land ownership and landscape-level factors on rare-species richness in natural areas of southern Ontario, Canada. Landscape Ecology 18:621–633.

Luther, D., J. Hilty, J. Weiss, C. Cornwall, M. Wipf, and G. Ballard. 2008. Assessing the impact of local habitat variables and landscape context on riparian birds in agricultural, urbanized, and native landscapes. Biodiversity and Conservation 17(8):1923–1935.

Madge, D. 2005. Organic viticulture: An Australian manual. Department of Primary Industries, Primary Industries Research Victoria, Irymple, Victoria. Australian Government/Grape and Wine Research and Development Corporation. www.dpi. vic.gov.au.

Maezono, Y., and T. Miyashita. 2003. Community-level impacts induced by introduced largemouth bass and bluegill in farm ponds in Japan. Biological Conservation 109:111–121.

McGourty, G., J. Nosera, S. Tylicki, and A. Toth. 2008. Self-reseeding annual legumes evaluated as cover crops for untilled vineyards. California Agriculture 54(3):191–194.

Merckx, T., H. Van Dyck, B. Karlsson, and O. Leimar. 2003. The evolution of movements and behaviour at boundaries in different landscapes: A common arena experiment with butterflies. Proceedings of The Royal Society of London, Series B-Biological Sciences 270(1526):1815–1821.

Merenlender, A. M. 2000. Mapping vineyard expansion provides information on agriculture and the environment. California Agriculture 54(3):7–12.

———. 2008. Collaborative conservation helps achieve regional water-quantity goals. California Agriculture 62(4):152.

Merenlender, A. M., M. J. Deitch, and S. Feirer. 2008. Decision support tool seeks to aid stream-flow recovery and enhance water security. California Agriculture 62(4):148–155.

Moonen, A. C., and P. Barberi. 2008. Functional biodiversity: An agroecosystem approach. Agriculture Ecosystems & Environment 127(1-2):7-21.

Noss, R. 1990. Indicators for monitoring biodiversity: A hierarchical approach. Conservation Biology 4:355–364.

Paoletti, M. G., and D. Pimentel (guest editors). 1992. Biotic diversity in agroecosystems. Special issue of Agriculture, Ecosystems, and Environment. 40.

Perkins, A. W., B. J. Johnson, and E. E. Blankenship. 2003. Response of riparian avifauna to percentage and pattern of woody cover in an agricultural landscape. Wildlife Society Bulletin 31:642–660.

Ruggiero, A., R. Céréghino, P. Marty, and S. Angélibert. 2008. Farm ponds make a contribution to the biodiversity of aquatic insects in a French agricultural landscape. C. R. Biologies 331:298–308.

Rustigian, H. L., M. V. Santelmann, and N. H. Schumaker. 2003. Assessing the potential impacts of alternative landscape designs on amphibian population dynamics. Landscape Ecology 18(1):65–81.

Santelmann, M. V., D. White, K. Freemark, J. I. Nassauer, J. M. Eilers, K. B. Vache, B. J. Danielson, R. C. Corry, M. E. Clark, S. Polasky, R. M. Cruse, J. Sifneos, H. Rustigian, C. Coiner, J. Wu, and D. Debinski. 2004. Assessing alternative futures for agriculture in Iowa, USA. Landscape Ecology 19(4):357–374.

Sax, D. F., and S. D. Gaines. 2003. Species diversity: From global decreases to local increases. Trends in Ecology and Evolution 18(11), November 2003.

———. 2008. Species invasions and extinction: The future of native biodiversity on islands. Proceedings of the National Academy of Sciences of the United States of America 105:11490–11497, Suppl. 1.

Semlitsch, R. D. 2000. Principles for management of aquatic-breeding amphibians. Journal of Wildlife Management 64:615–631.

Shutler, D., A. Mullie, and R. G. Clark. 2000. Bird communities of prairie uplands and wetlands in relation to farming practices in Saskatchewan. Conservation Biology 14:1441–1451.

Skorupa, J. P., and R. H. Hothem. 1985. Consumption of commercially grown grapes by American robins: A field evaluation of laboratory estimates. Journal of Field Ornithology 56(4):369–378.

Smart, D. R., E. Carlisle, M. Goebel, and B. A. Nunez. 2005. Transverse hydraulic redistribution by a grapevine. Plant Cell and Environment 28(2):157–166.

Smith, R., L. Bettiga, M. Cahn, K. Baumgartner, L. E. Jackson, and T. Bensen. 2008. Vineyard floor management affects soil, plant nutrition, and grape yield and quality. California Agriculture 54(3):184–190.

Sparling, D. W., G. M. Fellers, and L. L. McConnell. 2001. Pesticides and amphibian population declines in California, USA. Environmental Toxicology and Chemistry 20(7):1591–1595.

Stanley, E. H., and M. W. Doyle. 2003. Trading off: The ecological effects of dam removal. Frontiers in Ecology and the Environment 1:15–22.

Stephens, S. E., D. N. Koons, J. J. Rotella, and D. W. Willey. 2004. Effects of habitat fragmentation on avian nesting success: A review of the evidence at multiple spatial scales. Biological Conservation 115(1):101–110.

Sweet, R. M. 2006. Influence of cover crops on vine performance at two Willamette Valley vineyards. M.S. Thesis, Department of Horticulture, Oregon State University, Corvallis, Oregon.

Swolgaard C. A., K. A. Reeves, and D. A. Bell. 2008. Foraging by Swainson's hawks in a vineyard-dominated landscape. Journal of Raptor Research 42(3):188–196.

Tan, S., and G. D. Crabtree. 1990. Competition between perennial ryegrass sod and 'Chardonnay' wine grapes for mineral nutrients. HortScience 25:533–535.

Tobin, M. E. 1984. Relative grape damaging potential of three species of birds. California Agriculture 38(3,4):9–10.

Torchin, M. E., K. D. Lafferty, A. P. Dobson, V. J. McKenzie, and A. M. Kuris. 2003. Introduced species and their missing parasites. Nature 421:628–630.

Trenham, P. C. 2001. Terrestrial habitat use by adult California tiger salamanders. Journal of Herpetology 35:343–346.

Usher, M. B., and S. W. J. Keiller. 1998. The Macrolepidoptera of farm woodlands: Determinants of diversity and community structure. Biodiversity and Conservation 7(6): 725–748.

Van Elsen, T. 2000. Species diversity as a task for organic agriculture in Europe. Agriculture Ecosystems and Environment. 77:101–109.

Whisson, D. A., and G. A. Giusti. 1998. Vertebrate pests. In C. A. Ingels et al. (eds.), Cover cropping in vineyards: A grower's handbook. Oakland: University of California, Division of Agriculture and Natural Resources, Publication 3338. Pp. 126–131.

Wissmar, R. C. 2004. Riparian corridors of eastern Oregon and Washington: Functions and sustainability along lowland-arid to mountain gradients. Aquatic Science 66:373–387.

Wolpert, J. A., P. A. Philips, R. K. Striegler, M. V. McKenry, and J. H. Foot. 1993. Berber orchardgrass tested as cover crop in commercial vineyard. California Agriculture 47(5):23–25.

Young, T. P. 2000. Restoration ecology and conservation biology. Biological Conservation 92:73–83.

Zanini, F., A. Klingemann, R. Schlaepfer, and B. R. Schmidt. 2008. Landscape effects on anuran pond occupancy in an agricultural countryside: Barrier-based buffers predict distributions better than circular buffers. Canadian Journal of Zoology-Revue Canadienne de Zoologie 86(7):692–699.

PART 5

Organic Winegrowing Farming Practices

PHOTO: TOM LIDEN

Organic Viticultural Practices

GLENN T. MCGOURTY AND TOM PIPER

INTEGRATED CANOPY MANAGEMENT

Most winegrowers have the goal of producing high-quality fruit that will make wines that are distinctive, concentrated, and balanced in their flavors. Because organic winegrowers often choose to have lower yields than conventional growers, they are particularly focused on maximizing their grapes' quality. Organic producers use the standard viticultural practices employed by conventional growers as well as other, specialized practices. All winegrowers, and especially organic winegrowers, need to think about their canopy management practices systematically and take some time to evaluate how pruning, irrigation, fertility, cultivar and rootstock, their trellis system, and canopy manipulations affect production, vine vigor, wine quality, and profitability.

Anticipating Problems

Organic growers do not always have highly effective pesticides they can rely upon to stop an insect infestation or disease problem. Because of this, an organic grower's pest management strategy must focus much more on preventing problems before they occur. Well-balanced vines with an open, well-aerated fruit zone minimize the potential for powdery mildew and bunch rot problems in the fruit. Vines need to be properly irrigated, and this can be the single most important practice in the vineyard, particularly in warmer growing areas. Overirrigation promotes rampant growth, which in turn attracts leafhoppers. Underirrigation causes water stress, which make vines more attractive to spider mites. Water stress can also delay ripening and reduce fruit quality. Proper irrigation practices will help the vines to produce high-quality fruit and, ultimately, high-quality wine, free from stuck

fermentations and other problems in the winery. A vineyard where shoot positioning and leaf removal are completed in a timely manner is less likely to suffer sunburn of developing fruit and more likely to allow crop protectants to penetrate the canopy and prevent damage from pests and diseases.

Creating Vine Balance

Understand your site. Vineyard sites vary greatly, and it is important that your vineyard design and variety selection be site-specific if you are going to achieve good economic returns, grow quality fruit, and make good wine. Your choices of cultivar, rootstock, vine spacing, and trellis system have very long term effects on your vineyard, so you need to take your time to make these choices, using the best available knowledge from your potential customers, UC Farm Advisors and Extension Specialists, consultants, other growers, and grapevine nurseries. You will have opportunities from season to season to adjust vine vigor with fertilizer, irrigation, pruning, and vineyard floor management practices. Many organic winegrowers want to run sustainable operations, so you may want to design your vineyard to minimize any ongoing need for the input of additional off-site, energy-intensive resources.

In the coastal viticultural areas of California, many vineyards are planted in deep soil with a high water-holding capacity and moderate fertility. Generally, these sites are best for white cultivars, which are capable of producing moderate to large crops of high-quality fruit. Red cultivars tend to be less successful in these locations, but if you use a devigorating rootstock, competitive cover crops, and careful irrigation, you may be able to grow high-quality fruit from these vines as well. Hillside vineyards with shallower soils and less fertility are a better choice for red cultivars.

In the interior viticultural areas of California, both red and white winegrapes are grown in similar soil types. Vigor is often controlled by irrigation management, trellis design and pruning, crop load, and competing cover crops. Since the average price per ton tends to be lower than in coastal districts, growers usually have high-yielding vineyards as part of their profitability strategy: the lower per-ton value is offset by a larger per-acre volume of fruit.

Excessive vine vigor can predispose the organic vineyard to many problems:

- Lush succulent growth, favored by leafhoppers, can foster the development of large, damaging populations of these pests, too large to be easily controlled with organic pesticides.

- Large leaves, secondary shoots, and shading create a poorly ventilated environment that favors plant disease, including powdery mildew and Botrytis.

- Vigorous shoot expansion during bloom and an accumulation of nitrates can cause shatter when the flowers are pollinating.

- Heavy shade from the current season's growth reduces the number of fruitful buds the vine will produce the following year, reducing crop size and quality.

Before planting, you can predict the vine vigor for a vineyard based on the soil's available water-holding capacity. This can be determined in a soils laboratory, usually by testing the saturation percentage of the soil (Table 12.1).

It is important to note that under irrigated conditions with fertigation, sites with an otherwise low potential for vigor can be greatly invigorated when farmed conventionally. Organic growers do not have many options for inexpensive, water-soluble fertilizers. Additionally, most organic growers want their vineyards to reflect the natural terroir of their site and have found that wine quality is improved through the use of soil organic matter building techniques (see chapter 4, Soil Management in Organically Farmed Vineyards). Consequently, they try to match the vigor potential of their site rather than invigorating their vines with fertigation.

Assessing vigor. A grower can assess the vigor of a vineyard in several ways. Visual inspection during the growing season can reveal much information. Look for these features of a well-balanced vine:

- Ideal shoot length (normally between 36 and 42 inches, with 16 to 18 leaves per shoot).

- By veraison, shoots begin to lignify.

- Few lateral shoots (secondary shoots).

- Leaves are a healthy green color.

- Canopy appears open and well ventilated.

- Vine utilizes all of its allotted space in the trellis system.

- Shade beneath the canopy is dappled with sunlight, not solid shade.

Table 12.1. Relationship of saturation percentage to soil texture, cation-exchange capacity (CEC), and available water (field capacity – permanent wilting point)

Saturation	Soil texture	CEC (meq/100g of soil)	Available water (inches)	Potential grapevine vigor*
< 20%	sandy or sandy loam	2 to 7	< 0.6	very low
20 to 35%	sandy loam	7 to 15	0.6 to 1.0	low to moderate
35 to 50%	loam or silt loam	15 to 30	1.0 to 1.4	moderate to high
50 to 65%	clay loam	30 to 40	1.5 to 2.0	high to very high
> 65%	clay or peat	> 40	> 2.0	very high to extremely high

* Based on a 4-foot rooting depth with no chemical or physical rooting limitations. Deeper rooting depths or perched water tables may increase vigor.

- Two clusters of fruit are present on all shoots (possibly more if clusters are especially small—such as Pinot Gris—or fewer if clusters are especially large—such as Marsanne).

- Basal leaves (or the lowest leaves on the canes) are green and functional, not yellow or dry.

- Ripening fruit has uniform color from cluster to cluster and vine to vine.

Vine balance can also be evaluated on the basis of fruit-to-shoot ratios. This process is simple:

1. Select and flag 10 representative vines in each block at harvest.

2. Count the clusters and weigh the fruit on each of those vines.

3. When you return later to prune the dormant vines, count the shoots and weigh the prunings.

4. Calculate the ratio of the fruit weight to the pruned shoot weight. Table 12.2 notes the general relationships that are described in further detail below.

Reducing Vigor in an Organic Vineyard

If you are going to plant a new vineyard on a vigorous site that will be farmed organically, use the following strategies:

- Use rootstocks that do not invigorate.

- Choose cultivars that are not vigorous.

- Use a lower vine density and increase vine biomass (trunk, roots, canopy, and fruit).

- Consider more expansive trellis designs, especially those that allow shoots to "flop," since downward shoot positioning discourages vigor.

If you are farming an existing vineyard and want to reduce its vigor, follow these strategies:

- Use a cover crop strategy that will compete with vines for water and nutrients. Avoid cover crops that produce nitrogen. For example, use seed mixes that are mostly grasses and forbs and include only a small amount of legumes, if any.

- Use compost *only* as a soil amendment for the initial planting of the vineyard.

- Irrigate regularly to grow vines onto the trellis, and then use a regulated deficit irrigation (RDI) program (described later in this chapter).

- Do not apply fertilizers that contain nitrogen.

- Expand or modify the trellis system to better manage the canopy. This may allow you to increase the number of buds that you leave at pruning through the use of three-bud spurs, short kicker canes, an additional fruiting wire in cane-pruned systems, or other techniques.

Increasing Vigor in an Organic Vineyard

If you are going to plant a new vineyard on a low-vigor site that will be farmed organically, use these strategies:

- Use rootstocks that are vigorous.

- Choose varieties that are vigorous (if you have a choice).

Table 12.2. Fruit-to-shoot relationships and vine vigor

Mean cane weight	Ratio of fruit weight to weight of prunings	Vigor
> 60 grams	< 3:1	Vines overly vigorous. Too much wood, not enough fruit. Wine often has "veggie" flavors. Shading of fruit zone may cause poor flower initiation.
20 to 40 grams	4:1 to 6:1	Vines are balanced, especially at 5:1. Wines have good balance of alcohol, acidity, tannins and color.
< 10 grams	> 7:1	Vines are overcropped, not vigorous. More canopy is needed to properly ripen fruit. Wines tend to be light in color, with low tannins, high pH, and low acidity.

From Smart and Robinson 1992.

THE COST OF ORGANIC WINEGROWING

Organic winegrowers generally believe that there is little difference between the cost of organic and conventional winegrowing. There is an initial investment required for specialized equipment, such as under-the-vine cultivators for weed control, compost spreaders, and seed drills, but many conventional growers also purchase this equipment to improve their sustainable practices. Organic growers usually spend less money on crop protectants, since they have no herbicides or expensive insecticides and fungicides to purchase. Labor requirements for the two systems are similar for most operations (pruning, canopy management, harvest), except that organic winegrowers often spend more money on hand weeding since most mechanical under-the-vine cultivators do not control all weeds, especially around end posts and vine trunks. Additional hand weeding is usually done for aesthetic reasons rather than to reduce weed competition that might be detrimental to the vines.

Some wineries will pay a premium for certified organic winegrapes if they are pursuing the organic wine market. Most wineries pay based on the overall quality of the fruit, and prices are set based on the price point of the market niche that they anticipate to target with their wine. A simple formula is that the bottle price × 100 = the approximate price per ton that a winery will pay for the fruit.

Yields for organically farmed vineyards are similar to conventional vineyards yields, although growers may find it more of a challenge to farm large crops of less expensive fruit (as is typical in the San Joaquin Valley) without using fungicides to prevent bunch rot. Maintaining adequate soil fertility under high-yielding conditions can also be a challenge for organic growers and may require special attention.

Grower experiences in the North Coast area over the last 20 years have not shown organic winegrowing to be inherently any more risky than conventional winegrowing. There has not been any crop failure that can be attributed to organic production practices. The weather (especially frost and rain) is more likely than production practices to be a factor in crop losses.

For detailed cost studies, please visit the UC Davis Department of Agricultural and Resource Economics website: http://www.agecon.ucdavis.edu.

- Use close spacing in the vine rows and between vine rows.

- Use a trellis system with vertical shoot positioning, which will increase vine vigor.

If you are farming an existing vineyard and want to increase its vigor, use these strategies:

- Try to determine why the vineyard is not vigorous. Make sure that there is no significant root impairment caused by nematodes or phylloxera. Test both the vineyard soil and plant tissue from the vines at appropriate times.

- Water more frequently.

- Increase compost application rates. Use compost with a low carbon-to-nitrogen (C:N) ratio (between 10:1 and 20:1).

- Apply fertilizers as needed to stimulate growth (but do not promote excessive green growth late in the season).

- Stimulate growth by decreasing the number of buds you retain at pruning.

CANOPY MANAGEMENT PRACTICES

Organic growers are very dependent on canopy management practices that help them control light penetration, provide aeration to prevent mildew, improve spray penetration, and improve vine balance. Recognize that many of these procedures are labor intensive and therefore expensive. If your vineyard's crop value is not very high, you may not want to make much use of these procedures. The following manipulations are common:

- Suckering the trunk and removing sterile shoots on the cordons, or from the interior of head-pruned vines.

- Leaf removal near to fruit set, when you remove the two basal leaves from around the flower clusters on the cooler or shadier part of the vine.

- Positioning shoots with movable foliage wires (if present in the trellis system).

- Light hedging to improve air movement and allow equipment to move through the vineyard without becoming entangled in shoots. Shoots should be at least 3 feet long after hedging;

otherwise, fruit ripening may be delayed. Hedging should be delayed as long as possible to prevent excessive lateral shoot growth. If you need to do multiple hedging, excessive vigor is a problem in your vineyard and you need to implement a plan to address this. That plan might include less watering, growing competitive cover crops, increasing the vines' fruit load (but not overcropping), or alteration of the trellis system to reduce vigor.

- Fruit thinning might be needed if vines are not making adequate growth. A general rule of thumb is that shoots should be at least 2 feet long for a single cluster and at least 3 feet long for two clusters. Shoots less than 2 feet long should have the fruit removed, especially if it is a young vine.

Irrigation

Irrigation is a normal practice in many California vineyards, although some vineyards with deep fertile soils rarely need to be irrigated, particularly in cooler areas with high rainfall. Overhead sprinklers are used where water is plentiful and frost protection is needed. Sprinkler irrigation can cause rot problems as the fruit matures, so you have to schedule irrigations to take place before ripening to avoid damage to grape clusters. The last irrigation usually is applied at veraison, just as the fruit is beginning to soften.

Increasing numbers of vineyards with sprinkler systems have drip systems as well. A drip irrigation system can place the water precisely beneath the vines and avoid wetting the fruit so that water can be applied right up until harvest. Sprinklers are commonly run long enough to apply 2 or 3 inches of water per irrigation set one or more times per growing season, depending on that season's evapotranspiration rate.

Regulated deficit irrigation (RDI) is a strategy in which the grower applies less irrigation than normal to the vines. RDI is not used on a young vineyard until it has fully expanded and filled its trellis system, usually after four or five growing seasons. To begin the RDI program, make sure the vineyard soil profile is fully charged with water at budbreak. In many vineyards, rainfall is sufficient to accomplish this most years. In drier climates, though, preseason irrigation may be necessary. Normally, water is then withheld while the vine

canopy is actively elongating and setting fruit, thus reducing vegetative growth and crop load. Initial vine growth is supported by stored soil moisture and later by applications of water to ensure little or no vine water stress. Once the vine reaches the desired level of growth, irrigation is applied. Often when RDI is used properly, grape clusters are lighter and looser, which can improve wine quality and reduce the potential for bunch rot. Canopies tend to be more open, and red fruit often colors well under these conditions. However, if you use RDI without paying close attention to the condition and needs of the vineyard, you can cause serious water stress problems. Before embarking on an RDI program, ensure that your irrigation system is functioning uniformly and properly. You may also want to take soil moisture measurements to help you judge vine water status.

When you use RDI, you measure a deficit threshold (a predetermined level of midday water deficit) using a pressure chamber (pressure bomb) to measure vine leaf water status. The usual practice is to choose a fully expanded leaf growing in the sunshine, detach it from the vine, cut the end of its petiole, and place the leaf in a plastic bag and then into the pressure bomb with the cut end of the petiole sticking out. The chamber is then pressurized until sap is exuded from the end of the petiole. The amount of pressure needed to move sap from the leaf tells you how tightly moisture is being held by the leaf. This is measured in bars or atmospheres. Most grapevines begin to show stress at –12 bars; others may not show stress until –15 bars. Generally, growers with white wine cultivars or cultivars from northern climates (such as Pinot Noir and Merlot) use –12 bars as the threshold to begin irrigation. Red cultivars from warmer climates (such as Zinfandel, Syrah, Mourvedre, and Sangiovese) can use an irrigation threshold of –15 bars. Visual inspection should confirm that shoot tip growth is slowing, showing small internodes and small tendrils.

Once irrigation begins, many growers apply an amount of water based on a calculation of the percentage of the ground that is shaded by the vineyard canopy multiplied by the evapotranspiration (ET) rates for a given period of time. For instance, if ET for the week were 15 gallons of water per vine but the vineyard canopy only covered 60 percent of the vineyard floor, 9 gallons per vine would be applied (15 gal × 0.60 = 9 gal).

This topic is well covered in *Winegrape Irrigation Scheduling Using Deficit Irrigation Techniques* (Prichard 2003), which is available online and should be consulted for more details. Many organic winegrowers use this process to conserve water, improve fruit quality, and minimize conditions that encourage fungal growth. RDI also helps with vine balance, especially if irrigation practices in the past have led to overly vigorous vines.

Finally, you can continue to irrigate right up until harvest, provided that vine shoot tip growth has stopped. It is important to keep the canopy in good condition so the leaves will photosynthesize and sugar will accumulate in the fruit. If sugar maturity is coming too soon, you can slow the ripening process by applying extra irrigations and allow more time for tannins to mature and fruit flavors to develop.

Calendar of Organic Vineyard Practices

October and November:

- Harvest is completed.

- Cover crops are seeded, if necessary. Seedbed preparation usually involves tillage, although no-till seeders (slit seeders) can be effective.

- Apply compost, lime or gypsum, nitrogen, potassium, boron, and phosphorus fertilizers as needed.

- Postharvest irrigation, applied if the vines are actively growing, especially for earlier-ripening cultivars such as Chardonnay and Sauvignon Blanc. In warmer and drier areas, it is important to re-wet the root zone and leach salts.

December and January:

- Begin pruning on cultivars that are not highly susceptible to Eutypa.

- Pay crop assessment to certifiers.

- Arrange for inspector visits and an audit of your organic farm plan. Review the organic plan for the vineyard and make changes if needed.

February and March:

- Complete pruning.

- Tie vines.

- Start your beneath-the-vine weed control program if you are tilling (weather permitting).

- Inspect and pre-test your frost-protection system.

- At budbreak, apply necessary crop protectants (wettable sulfur or stylet oil) to control overwintering mites and cleistothecia of powdery mildew.

- Prepare for frost protection; obtain fruit frost forecasts if available for your area.

- If in Pierce's disease area, consider applying kaolinitic clay to protect vines from sharpshooter damage and feeding as new foliage emerges on vines.

- Begin sheep grazing, if that is part of your plan. Avoid grazing if the vineyard soil is wet or likely to be compacted. Remove animals when buds begin to break.

- Meet with your winemaker to review last season's wines. Taste tank or barrel wine samples if possible. Discuss possible changes or improvements in winegrowing practices.

April and May:

- Apply vineyard floor management practices, such as mowing and incorporation of cover crops, if needed. Begin or continue sheep grazing, if planned, before budbreak.

- Continue beneath-the-vine tillage for weed control.

- Canopy management practices, including suckering and shoot positioning.

- Continue frost protection, if needed.

- Take petiole samples at bloom time.

- Apply pre-bloom foliar fertilizers (containing zinc and boron) if needed.

- Continue crop protectants for powdery mildew control.

- Monitor for vineyard pests such as leafhoppers and mites.

June and July:

- Vineyard floor management practices such as mowing and incorporation of cover crops, if needed.

- Canopy management practices, including leaf removal, shoot positioning, and hedging.

- Begin irrigation, either on the basis of visual determination or a check of vine moisture status using a pressure bomb or a soil moisture sensing device.

- Apply crop protectants for powdery mildew control. Do not use wettable sulfur after bloom. Do not use sulfur dust when temperatures will exceed 95°F.

- Monitor for vineyard pests such as leafhoppers and mites. Consider spraying if pest populations approach economic thresholds and beneficial insects are not sufficient to control the problem.

August and September:

- Continue irrigation as needed. If fruit is ripening too quickly, apply extra irrigation to slow the accumulation of sugar. If fruit ripening is delayed, continue regulated deficit irrigation.

- Thin the fruit before or at veraison, if needed. Remove any fruit that are not coloring uniformly.

- Place bird netting on the vines at veraison, if needed.

- Monitor for vineyard pests such as leafhoppers and mites. Consider spraying if pest populations approach economic thresholds and beneficial insects are not sufficient to control the problem.

- Begin sugar testing of fruit 2 weeks after veraison.

- Prepare equipment for harvest.

- Make arrangements for pickers and transportation of fruit to the winery.

- Contact the winery when sugar tests are close to specified target levels for harvest. Consult with your winemaker to determine the picking time and date.

- Harvest begins.

- Order cover crop seeds.

REFERENCES/RESOURCES

Prichard, T. 2003. Winegrape irrigation scheduling using deficit irrigation techniques. Davis: Land and Water Reources, University of California, Davis. http://ucce.ucdavis.edu/files/filelibrary/2019/13563.pdf.

Smart, R., and M. Robinson. 1992. Sunlight into wine: A handbook for winegrape canopy management. Ministry of Agriculture and Fisheries, New Zealand.

Winkler, A. J., J. Cook, W. M. Kliewer, and L. Lider. 1974. General viticulture. Berkeley: University of California Press.

INDEX

Note: Page numbers in **bold type** indicate major discussions. *Italic* type is used to indicate tables, e.g., *86t,* and figures, e.g., *115f.*